The Last Two

The Last Two

The Battle to Save the Northern White Rhinos

Boštjan Videmšek and Maja Prijatelj Videmšek

Photographs by Matjaž Krivic

ROWMAN & LITTLEFIELD
Lanham • Boulder • New York • London

All photos © Matjaž Krivic.

Published by Rowman & Littlefield
An imprint of The Rowman & Littlefield Publishing Group, Inc.
4501 Forbes Boulevard, Suite 200, Lanham, Maryland 20706
www.rowman.com

86-90 Paul Street, London EC2A 4NE

British Library Cataloguing in Publication Information Available

Library of Congress Cataloging-in-Publication Data

978-1-5381-7846-1 (cloth)
978-1-5381-7847-8 (electronic)

For David Beriain (1977–2021), a twin brother

We are not an endangered species ourselves yet, but this is not for lack of trying.

—Douglas Adams, *Last Chance to See*[1]

Contents

Acknowledgments

We are grateful to all the staff in Ol Pejeta Conservancy for giving us an opportunity to spend numerous mornings with Najin and Fatu, especially their main caretaker Zachary Mutai.

We would also like to thank the whole BioRescue team for all their help and support. We are extremely grateful to the rangers in Samburu National Park who took us on board, and the team of CROW instructors.

We would also like to express our gratitude to our Slovenian publisher UMCo and its editor Samo Rugelj—he was the first to recognize the value of this story and published the Slovenian edition of the book in 2022.

Last but not least, we say thank you to our U.S. editor Deni Remsberg, who has recognized this book's international potential and smoothly led us through the editorial process.

INTRODUCTION
Waiting for a Color Postcard

We moved from being a part of nature to being apart from nature.
—David Attenborough[1]

A BLACK-AND-WHITE POSTCARD FROM THE DVŮR KRÁLOVÉ ZOO IN the Czech Republic bears the image of the last two northern white rhinos in existence. An inscription at the bottom says: "Fatu, the young millennial." The year when Fatu was born, 2000, our planet saw the birth of 132 million children and a single northern white rhino. . . . Probably the last of its kind.

Najin, Fatu's mother, was the penultimate northern white rhino born in captivity.

The postcard shows Najin and Fatu feeding on watermelons, apples, oranges, cabbages, and sunflowers. Both their domicile—hard ground, bereft of all greenery—and their calorie-rich meals were poor substitutes for what used to be the life of the northern white rhinos inhabiting the grassy savannahs of Central and Eastern Africa.

Najin frontally faces the camera lens. One's attention is immediately drawn to the way her longer horn curves down instead of up, the result of several shortenings. Zoos across the world regularly shorten the horns to prevent rhinos from hurting each other. In some African national parks and private reserves, the horns are also shortened to make the rhinos unattractive to poachers—or because the reserve owners are banking on the horn trade someday being deregulated.

The horns are the northern white rhinos' main means of defense. They are also their greatest curse.

Despite Najin clearly relishing the food, her not-so-sharp yet still highly alert gaze is directed at the photographer, a human being—her kind's one great natural enemy. Her body language clearly states: '*Do not approach. And above all, do not approach me.*' Meanwhile, Fatu—a small copy of her mother with the barest hint of a horn—confidently munches on her meal.

In 2000, when Fatu was born, the planet was home to six billion people. Now in 2023, there are eight billion of us. Fatu remains the last northern white rhino born in captivity, or anywhere at all.

By the time they reach her age, many of her human counterparts will have already produced offspring. Fatu, however, will never be able to do so, at least not in the natural way. The same goes for her 33-year-old mother.

February 2022, a Second Photo

Just after daybreak, Najin, Fatu, and their playmate—Tauwo the southern white rhino—left their heavily guarded pen under the acacia trees. They were headed off to graze at a nearby paddock at the Ol Pejeta Conservancy.

The Conservancy, located in central Kenya, has been the home of the world's last two female northern white rhinos since 2009. That year, they were brought here from the Dvůr Králové zoo—along with two males named Sudan and Suni. The hope was that the call of nature might persuade the four of them to mate.

As we stood and watched, the sun slowly rose above Mount Kenya. It painted the three rhinos' flanks in a rather magnificent golden-gray. Yet Najin, Fatu, and Tauwo paid no heed. Heads pointed to the ground, they slowly trundled around the grazing grounds and feasted on fresh grass. Their silhouettes stood out from the savannah plains like three mighty landscapes of their own, three lone islands dotting a grassy ocean.

Some experts classify the northern white rhinos as a subspecies of the southern white rhinos. Others prefer to see them as a wholly separate species.

The northern variety is somewhat smaller and lighter than its southern peer. The typical northern white rhino weighs between 1.4 and 1.6 tons, whereas the typical southern white rhino comes in at between 1.7 and 2.7 tons. The northern white rhino is also marked by the concave shape of their heads, which it keeps somewhat lower to the ground. The southern white rhino has a flatter and more upward-thrusting head. The back of the northern white rhino is flatter. Its ears and tail are furrier, and it also boasts a longer tail.

Yet both the northern and the southern variety are marked by an even upper lip. This is an evolutionary advantage when it comes to grazing. It separates them from the black rhinos, whose more hook-shaped lips help them easily pick off leaves from bushes and trees.

As she grazed herself to her heart's content, Najin was easily identified by her long upward-curving horn, her sagging gait, and a constant companion—her caretaker and best friend Zachary Mutai. The slightly built man kept stroking her, removed ticks, and scanned for patches of cracked skin. If left unsupervised, these could lead to inflammation and serious trouble.

Najin was clearly the star of Zachary's own personal movie.

"She's my second family," he smiled in the morning sun. His mission in life is to look after her until the very end. Just like he looked after Sudan, the last northern white rhino male, who died of natural causes at the age of 45 in 2018.

With Sudan's demise, the northern white rhinos have become essentially extinct.

The last spotting of their kind in the wild took place in 2006, at the Garamba National Park in the Democratic Republic of Congo. A year later, some fresh tracks were found nearby. This marked the last human contact with the northern white rhino in the wild.

For a while, there was hope that a lone embattled member of the species might still be out there—perhaps in Sudan, or in the DRC—

toughing it out in the brushes like some poor maddened member of the Viet Cong.

These hopes are long gone.

Even if it still existed, the surviving northern white rhino would have been found by poachers and would have been mercilessly butchered. First, he or she would have been felled by automatic rifles or poisoned darts. Then the horns would have been chopped off, probably along with a large chunk of the head.

Much like our own nails and hair, rhino horns are made of keratin. The protein giving firmness to these structures has no scientifically proven medicinal value whatsoever. Yet for thousands of years, practitioners of Chinese traditional medicine have been ascribing it all but magical properties. This is the main reason why rhino horns are so highly sought after by international crime syndicates.

Over the last 25 years, China and Vietnam have established themselves as the planet's two largest markets for rhino-horn products by a long shot. The organizations dominating the illicit horn trade are therefore mostly based in Asia. Most of them can rely on substantial support from the security structures of the countries where rhinos are being slaughtered. Along with southern and northern white rhinos, the hardest hit were the black rhinos.

Somewhere down the path toward the end user, the horns are ground into a fine powder. This powder, mistaken for a cure-all for at least 2,200 years, then finds its way on to the black market, where it is touted as a cure for _____—fill in the blank with whatever ails you.

The substance, for example, remains widely marketed as an aphrodisiac and a hangover antidote. It has also become quite a status symbol. At the time of writing, one kilogram of rhino-horn powder fetches between $60,000 and $100,000 on the Asian market. That makes it pricier than either heroin or cocaine.

Delusion, deception, and greed have always formed a key part of human civilization. They continue to do so now. Rhinos, of course, are not the only victims.

There Is Also a Third Photo

Perhaps it would be better to call it an image. It is the digital version of an old map depicting the northern white rhino's natural habitat.

To anyone with some basic understanding of recent history, this map offers the fastest possible explanation as to why the species is now on the very brink of extinction.

Over the last few decades, the northern white rhino had the misfortune of being caught up in all the key catastrophes of Central and Eastern Africa. This separates the species from its closest cousin, the southern white rhino, which used to be critically endangered but has so far managed to survive. Various relocation programs, poaching bans, and breeding programs proved sufficient to keep them struggling on.

The northern white rhino, on the other hand, was not so fortunate. Its all-but-certain extinction was greatly expedited by a number of wars raging across their natural habitats. There was the civil war in Chad, The 'African World War' in the Democratic Republic of the Congo, the genocide in Rwanda, the war in Burundi, several wars in Sudan and South Sudan, the aftershocks of the civil war in Somalia, and the chaos in Uganda.

If we were to superimpose the map's two key elements—namely the African wars between 1980 and 2010 and the living space of the northern white rhinos—we would get an almost perfect match. In more ways than one, these rhinos were *prisoners of geography*, cursed by where they happened to live.

Cursed and condemned to extinction.

The wars, along with the wholesale devastation that they wrought on the rhinos' natural habitats, opened the door wide to poachers. Some of them saw an opportunity to make a quick buck, and others took to killing so they themselves might survive. Their reasons do not matter all that much now. The grim fact is that the northern white rhino was endangered long before the blood-soaked finale of the twentieth century. Precious few good souls were on hand to stand with them during the *rhinocide* perpetrated by increasingly emboldened poachers.

The butchery took place on an industrial scale. The process was so fast and thorough that the conservation community was unable to respond with anything resembling adequate countermeasures. Shoddy legislation, endless political machinations and rampant local corruption were just some of the causes that led up to the current disaster.

From 2007 on, the northern white rhino could only be found in two of the world's zoos: the Czech Dvůr Králové Zoo and the San Diego Zoo in California. Even there, they were failing to reproduce. It was clear something had to be done or the species would forever be lost.

2009 was the year when the management of the unique Dvůr Králové Zoo—home to more southern white rhinos and black rhinos than anywhere else in the world outside Africa—was approached by an international team of scientists and conservationists. Their proposal was to relocate four of the zoo's northern white rhinos to central Kenya.

And so it happened that Sudan, Suni, Najin, and Fatu were transported to the Ol Pejeta Conservancy. They may have been used to their caretakers' commands being issued in the Czech language, but the four new arrivals seemed to adapt reasonably well to life in Kenya.

No offspring, however, seemed to be in the cards. Given Najin and Fatu's gynecological problems, it was understood that a pregnancy was all but impossible. Barring a miracle, nothing could be done about the species' imminent extinction.

Then it turned out that maybe—just maybe—a miracle could be provided by modern science.

In 2015, the BioRescue international consortium was founded in Vienna, Austria. The organization's mission is to save functionally extinct species from final extinction.

Among the consortium's founding members were:

- The Safari Park Dvůr Králové Zoo
- The Leibniz Institute for Zoo and Wildlife Research
- The University of Padua
- Avantea, a Cremona-based company specializing in reproductive technology
- The Max Delbrück Center for Molecular Medicine from Berlin
- The Kyushu University from Japan
- The Ol Osaka Pejeta Conservancy from Kenya
- The Merck Company
- Kenya Wildlife Service, the Kenyan state institution dedicated to wildlife conservation

The first pillar on which the project rests is the collection of so-called "oocytes," or the female rhino's immature eggs. Once secured, the oocytes are to be fertilized with the sperm of deceased males, and the resultant embryo transplanted into a southern-white surrogate mother. The second pillar that gives hope to the conservationists is oocyte creation from the skin cells of northern white rhinos, with the help of stem-cell technologies.

The days Najin and Fatu spend in their luxurious refugee camp—outside of it, they would be immediately killed and butchered—largely resemble each other.

Early to rise, the last two of their kind normally graze until noon. They follow a long visit to the water trough by resting in the shade, then it is time for an afternoon's worth of grazing and an early night. During the rainy seasons, this routine is enlivened with regular mud baths, providing tremendous enjoyment to the two animals.

In many ways, Najin and Fatu reside in rhino heaven. Blissful unawareness of their species' doom is only one of their blessings.

On the other hand, the long years of agonizing over the prospects of saving the northern white rhinos have turned many of the BioRescue scientists' hair gray. These researchers and conservationists are waging a war against time, with no room for error. The technologies at their disposal have never been used on rhinos—or on any other animals, for that matter.

It is such a simple thing to take an animal life. Our entire civilization, after all, has been built on the subjugation of the animal world. On the other hand, to create life—or at least preserve it—is proving an infinitely more difficult task.

In late 2021, Najin started living the life of a privileged pensioner. Her advanced age and medical problems finally excused her from the egg-collecting program, which involved putting her to sleep every three months. The combination of Najin's eggs and the sperm of deceased males failed to produce a single embryo. All the ones obtained so far, twenty-two in all, have been created from Fatu's eggs.

Fatu, however, will be unable to carry any of the embryos. The reason is her endometriosis, a chronic disease-causing womb-like tissue to grow outside of the womb and resulting in an inability to get pregnant.

Therefore, the tiny bundle of hope will be transferred into a surrogate mother—one of the southern white rhinos already residing at Ol Pejeta.

The project's main driving force, the German veterinarian Thomas Hildebrandt, wants to initiate the transfer as soon as possible. If all goes according to plan, the next baby northern white would be born approximately 16 months later.

Accompanied by caretaker Zachary Mutai, a good man if ever there was one, Najin and Fatu shambled from their pen toward the savannah, as the cold mountain wind picked up.

The last two's current refuge is in the Kenyan province of Laikipia, almost 2,000 meters above sea level. Zachary, who dislikes heavy clothing, trembled from the early morning chill. The two of us were completely unprepared for winter conditions in the middle of Kenya. We kept marveling at our chattering teeth. Yet we took much warmth and comfort from both majestic and vulnerable radiance exuded by our two wards.

All our research had made us appreciate how vicious and all-devouring the storm must have been—the storm that had turned Najin and Fatu into the last two members of their iconic species.

We took courage from the fact that they were still very much alive and that they were an integral part of a fascinating story in the making, a story that involves each and every one of us.

Should BioRescue's endeavors prove successful, there is a chance for the functionally extinct species to survive. Granted, saving one species hardly means saving the world. But if scientists can perfect the technologies for one species' salvation, these technologies could then be used for saving many others. Of course, that will only matter if the rest of us somehow break out of our murdering and pillaging algorithms.

Then, and only then, there is still hope.

That is really what we are down to. The forces behind the extinction of the northern white rhino and thousands of other species are the exact same forces behind climate change.

They are the greatest existing threat to our children's survival.

We all need something to live for, a spiritual base from which to push back against all the apathy, cynicism, and nihilism tearing the planet apart. The elusive "something" being invoked here is itself an endangered species. There may not even be a proper word for it, as things stand. And so, we all need to invent it: a word denoting respect, courage, moderation, solidarity, mindfulness. . . . And above all, a dedication to change.

Maybe, therefore, the two of us so quickly found peace immersing ourselves in the project of saving the northern white rhino. What follows is a tale of love. Najin and Fatu's first gift to us was some much-needed respite from the horrors of current journalism.

And then their gifts just kept coming.

On that fateful daybreak when we first looked into the moist and sleepy eyes of the two astonishingly beautiful animals, we immediately knew that we wanted to write a book about them. Together, we wanted—even needed—to tell their tale.

Why?

This is why: "It's essential now, as the prospect of planetary catastrophe comes ever closer, that those nonhuman voices be restored to our stories. The fate of humans, all our relatives, depends on it."[2]

The quote comes from *The Nutmeg's Curse: Parables for a Planet in Crisis*. With each passing day, the book offers a more painfully pertinent read. The author, Amitav Ghosh, is banking on our own species' ability to tell stories. "This is the big burden that now rests upon writers, artists, filmmakers, and everyone who is involved in the telling of stories: to us falls the task of imaginatively restoring agency and voice to nonhumans." In Ghosh's view, this is one of the key missions of our time, given the extent of the planetary crisis.

So, this is the story of the last two northern white rhinos: the story of how Najin and Fatu got bereft of their kin, and how some unusually well-meaning people got together to try to turn the murderous tide.

Undaunted by the seemingly insoluble riddle of resurrecting a functionally extinct species, the scientists involved now say they may very well be close to solving it. During this book's final revision, it was almost time for the first embryo to be transferred into a southern-white surrogate mother.

All we can do now is to wait for a full-color postcard of Fatu and a baby with the barest hint of a horn. In this snapshot, it would be Fatu's job to glare at the photographer, her body language stating: "Do not approach!"

The northern white rhinos *must* survive, and with it, the hope that our own kind is still capable of doing good, of reversing the trend—or at least of starting to mitigate the destruction that our rise has wrought on our surroundings.

The northern white rhino must survive because hope must survive. So yes, in this one sense, saving one species can indeed mean saving the world—and with it, ourselves.

As lifelong atheists, we both can't seem to recall the last time we prayed. But we are praying now.

The Rhinocide

The home of researcher and legendary conservationist Kes Hillman-Smith is located at the outskirts of Nairobi. It is chock full of rhinos.

The shelves are crowded with statuettes of every imaginable shape and size. The walls are covered with rhino images—some of them photographs, some drawn by our hostess's children and grandchildren. A huge rhinoceros adorns the buckle of the wide belt holding Kes's jeans in place. A rhino from the Garamba National Park in the Democratic Republic of Congo is the centerpiece of a remarkable stained-glass painting in the nursery room of her grown-up daughter.

The driveway to the Hillman-Smith estate may be decorated with a sign stating Hog Heaven, but the place itself is a shrine to the northern white rhinos.

Ceratotherium simum cottoni.

The scientific community learned of the species—or, according to some researchers, subspecies—in the spring of 1903, when a specimen killed near the Lado settlement along the banks of the White Nile River was put on exhibition. In 1908, the first scientific description was to follow.[1]

It was only much later that it was determined that the northern white rhino was quite plentiful, albeit confined to a surprisingly limited area. Most of them could be found in southern Chad, the northern and eastern parts of the Central African Republic, the southwest of South

Sudan, the northwest of the Democratic Republic of Congo, and the northwest of Uganda.

By the 1960s, only 2,360 northern white rhinos remained. Some 1,200 of them lived at the Garamba National Park in the Democratic Republic of Congo. By the beginning of the seventies, the number of the Garamba rhinos was down to 490. By 1984, there were only fifteen left.[2]

Following tremendous efforts by Kes Hillman-Smith and her fellow conservationists, the number rose to approximately thirty by 2004 as Hillman-Smith told us in the interview. But that year also marked the next—and final—wave of extermination, after which only four members of the species survived in the wild. If there were a few more, no one ever saw them.

2006 was the year that the last northern white rhino was spotted in the wild. In 2007, the last known tracks were found—a few footprints along with some fresh excrement.[3]

The International Union for Conservation of Nature (IUCN) last reviewed the species in 2020. Within the organization's Red List of Threatened Species, the northern white rhino was placed in the category of critically endangered species, probably extinct in the wild. There is no doubt they are extinct in the Central African Republic, Chad, Sudan, and Uganda—and very probably also in the Democratic Republic of Congo as well as South Sudan.

From 1948 to the mid-seventies, twenty-two northern white rhinos were captured and transported to various zoos in Czechoslovakia, England, the Netherlands, the United States, Sudan, Saudi Arabia, and the United Arab Emirates. Unfortunately, the rhinos that were relocated only managed to reproduce at the Dvůr Králové Zoo in Czechoslovakia.

Four new members of the species were born in captivity: Suni, Nabire, Najin, and Fatu. The first three were delivered by Nasima, born around 1965 in Uganda. Then, in 2000, Najin gave birth to Fatu, the last northern white rhino to be born in captivity to date.

In 2010, only ten northern white rhinos were still alive, all of them in captivity.

2014 marked the death of Suni at the OI Pejeta Conservancy and Angalifu at the San Diego Zoo.

2015 saw the passing of Nola at the San Diego Zoo and Nabire at the Safari Park in Dvůr Králové).

In 2018, Sudan—the last of the northern white rhino males—died at the Ol Pejeta Conservancy.

Today, only Najin and Fatu are left.

Kes Hillman-Smith finds it very hard to shake the last vestiges of hope that some other specimen might be hiding somewhere out in the wild. Even if, after all this time, such a sighting would be a miracle.

WITNESSING A GREAT CRIME: KES HILLMAN-SMITH AND THE GARAMBA NATIONAL PARK

"At the Garamba, all we left behind were piles of smashed brick and a huge lemon tree. It sprung up from the lemon seeds we used to spit on the ground during breakfast. That tree, I guess, is what our legacy over there amounts to."

At the time of our visit, Kes Hillman-Smith was seventy-two. She was justly considered one of the century's greatest conservation heroes. But her tone, as she told the tale, was decidedly bitter, and her sun-and-pain-worn face seemed tired and grim.

This particular memory took her back to the 1990s when the northern part of the Democratic Republic of Congo was ravaged by several horrific wars. At the Garamba National Park in the northwest of the deeply traumatized country, Kes and her husband Fraser Smith were taking care of the last thirty or so northern white rhinos still living in their natural environment.

The Garamba, one of UNESCO's World Heritage Sites, is the second oldest African national park. It was founded in 1938, during Belgian colonial reign. Kes's tenure lasted for twenty-two years, from 1983 to 2005. This was a period of unconscionable brutality, one that left a profound mark on the history of the Congo and the African Great Lakes region.

Kes and her family lived through the First Congo War (1996–1997), which tore the country apart. Then they endured the Second Congolese War, the so-called "African world war," which killed six million people between 1998 and 2003: approximately the same number as the Holocaust.

Kes also bore witness to the ravages of genocide in neighboring Rwanda, where 800,000 people got massacred over two months in 1994.[4] The porous borders, countless refugee convoys, and the constant influx of warlords and mass murderers made sure the consequences of the industrial-scale butchery were very much felt in the Congo. They are still being felt to this day.

In addition, there was the fallout from the civil war in Sudan and the genocide in Darfur. The fiends so heedlessly murdering human beings hardly batted an eyelash while slaughtering animals by the truckful. Elephants and the rhinos were the hardest hit of all. That is hardly surprising, given the financial gain their demise brought to the war criminals and their foreign employers.

The tragic fact is that much of Africa's modern bloodshed was financed through the ivory and rhino-horn trades.

Kes Hillman-Smith comes from an English military family. Her father was a pilot with the British Royal Air Force. Her family moved so many times that she grew up unfamiliar with the concept of a permanent home. From her youngest years on, she found it very difficult to endure closed spaces. The only place she really felt at home was the sea.

In her own words, she is a woman of nature, of open horizons that promise freedom and independence. Along with her love of animals, this was the reason she chose to study biology at the University of Leicester. She eventually got a doctorate after doing her dissertation on pesticide harm to lemurs.

After finishing her education, she set out into the world. She viewed England as "the land of mediocrity," so she had no qualms about leaving it behind. While at the university, she was the president of the student travel society. She visited Norwegian Arctic, Romania, Turkey, and Iran. She travelled across much of Mongolia on horseback. But she had always wanted to strike off for Africa to make a life of her own.

In the first half of the 1970s, she decided to cross the Sahara Desert with a few friends. They spent six months in a Land Rover, driving through the desert from west to east. Her companions studied the handicraft of the local communities, while Kes photographed. They concluded the journey in Kenya, where they put up a handicraft exhibition. On arrival, Kes got word that her project of working with the common eland, a type of antelope, had been given the green light.

And so, she stayed in Kenya; her friends moved on.

She never regretted her decision. "In more ways than one, I fell in love," she told us, sitting on the verandah of her Kenyan home, which borders a local giraffe sanctuary.

The atmosphere pervading her estate was the opposite of the grotesquely overcrowded, suffocating, chaotic and visibly festering Kenyan capital. Hillman-Smith's verandah offered a splendid view of her garden, where the local wild boars were free to roam, beg for food, fight, and slumber. The tree affording shade to the verandah sported a bird feeder with sugared water, the source of much delight for the visiting sunbirds. They are like the Colibri birds, only less hectic.

The stone-tiled verandah also served as a landing strip for helmeted guinea fowls. Each day around five in the afternoon, when Kes served us a tray of strong black tea, dates and ginger cookies, several helmeted guinea fowl would land and start demanding dinner from our slight but only seemingly fragile hostess. At the same time, her dogs Soldier and Flea would pick up a markedly wolfish howl, signaling it was time for their constitutional stroll around the marvelous small forest at the giraffe sanctuary.

Among other things, the giraffe sanctuary also serves as a meeting place for Nairobi-based Europeans. They exchange news and enquire about each other's health, mutual acquaintances, and business ventures.

It is a world of its own.

"I met my first husband Chris Hillman standing over the body of an immobilized eland," Kes recounted as we accompanied her on the stroll, with Soldier and Flea tirelessly patrolling the thicket. "After the wedding,

I renounced my British citizenship to become a Kenyan, and I never looked back."

Africa and its northern white rhinos became her home, her love, and her destiny.

The whole continent, and Kenya in particular, made it impossible for her not to get involved with wildlife conservation. During the first half of the 1970s, many of her projects were dedicated to African elephants, who were then heavily targeted by poachers.

It was the biologist Iain Douglas-Hamilton who completed the first studies of elephant behavior in the African wild. He also developed methods for determining their overall numbers with the help of small aircraft. Between 1976 and 1979, a joint venture between The International Union for Conservation of Nature (IUCN) and The World Wildlife Fund for Nature (WWF) enabled him to make a comprehensive census of the African elephant across thirty-four countries. The main objective was to obtain precise data for forming guidelines to aid the elephant's conservation.[5]

The project's results revealed the enormous dimensions of the poaching problem that confronted the continent during the end of the 1970s and the first part of the 1980s: A crisis that was fueled by growing demand for ivory in Asia, especially Japan.

When Kes met the legendary biologist, the founder of the Save the Elephants organization and the author of several books, her fate was all but sealed.

She became Douglas-Hamilton's assistant. Their first joint mission was to count the elephants of Kenya from the air. "It was so infernally hot and chaotic. We kept circling and circling above the elephants and I was getting really sick," she recalled the unpleasant yet also seminal experience as we walked on.

The daughter of a military pilot had no problem securing a flying license of her own. Then, for a good long while, counting elephants and rhinos from above became the focus of her existence.

Iain Douglas-Hamilton, who is now one of her neighbors at the plateau above Nairobi, also collaborated with her on their great pan-African elephant census. Among other places, their sorties often took them over the

Garamba National Park's grounds. The bird's eye view of the savannah—its enchantingly tall grasses and river-strewn forests—was Kes's first contact with the place that was to become her home.

1978 was the year of the first African rhino census from the air. It was set up by Iain Douglas-Hamilton and led by Kes Hillman-Smith. The same project gave rise to the African Rhino Specialist Group (AfRSG). Kes served as its first president.

The group, which is still functioning today, started out as a voluntary association of experts striving to preserve the African rhino.

The rhinos were soon divided into three categories: black, southern white, and northern white. Being the most endangered, the northern white became the focus of the group's activities. The funds were provided by the Dutch branch of the World Wildlife Fund and by the controversial Dutch prince Bernhard, one of the WWF's founders.

"Even then, more than forty years ago, I knew that the rhinos were much more vulnerable than the elephants. Not that the elephants weren't highly vulnerable! But they were also much more plentiful. Both species were completely exposed to poaching and to legal hunting. Something had to be done!" Hillman-Smith related as we made halts on our forest stroll so she could point out a few local animals.

In the shadier part of the forest, we spotted a tree hyrax—a rabbit-sized forest animal—motionlessly clinging to a tree. If not for our hostess, we wouldn't have noticed it. When we reached the lowest point of the circular trail next to a nearly dried gulch, Kes whooped in delight and pointed at the brushes. On closer inspection, we could discern the glint of orange-blue feathers followed by a very long tail. It was the African paradise flycatcher male and its quiet splendor made for the highlight of our stroll.

At first, the idea was to set up a few protected areas for rhinos in Sudan, but civil war broke out there yet again. The African Rhino Specialist

Group thus switched focus to the Garamba National Park in the neighboring DR Congo, which still used to be Zaire back then.

At the time, this unique reserve for the African megafauna was where the northern white rhino was most plentiful.

"The Zaire authorities inquired whether I was available for Garamba. They wanted me to set up a monitoring program for elephants and rhinos. I told them: 'Maybe for a year'? Well, that was the start of my twenty-two years at Garamba."

In 1981, Kes successfully set up a research station and met her second husband, a ranger with the South African National Parks agency named Fraser Smith. They first crossed paths at a meeting of a task force created to transport black rhinos to Pilanesberg, a South-African national park.

"Fraser wanted to leave South Africa. He was a very practical man, extremely gifted in all technical matters. He was exactly the sort of person Garamba needed!" Hillman-Smith beamed as we reached a clearing with a mound of stones at one of its edges.

She picked up a small stone and added it to the stony pyramid with a deft throw. "This is a monument to one of the Maasai warrior chiefs," she explained. We were very honored to toss our own two pebbles onto the pile.

The three of us were silent for a few moments.

We thought: *Kes is a warrior, too.*

Will someone build a monument for her?

In 1983, Kes and Fraser completed the first airborne tally of northern white rhinos at Garamba and at Shambe National Park in the south of Sudan. The results were as far from encouraging as they could get.

As of 1960, 1,200 northern white rhinos were living at Garamba.[6] The latest tally pointed to a catastrophic reduction. At Shambe National Park, Kes and Fraser found only huge piles of rhino skeletons. At Garamba, they only managed to locate fifteen northern white rhinos. At the beginning of the 1970s, there were 500 of them.

The Hillman-Smiths were later able to determine that the most devastating causes were two separate poaching sprees.

Most of the northern white rhinos—as many as 80 percent of them—were killed during the Simba tribe's rebellion at the start of the 1960s, when the Belgian colonizers withdrew from the Congo. The second killing wave took place between 1978 and 1984, which meant that Kes and Fraser were making their tally while it still was going on.[7]

These findings were cause for great alarm. The two of them wasted no time setting up a constant monitoring and protection program for northern white rhinos at Garamba. In March 1984, they moved to the Congolese national park.

Kes Hillman-Smith's life mission had begun.

Kes and Fraser built themselves a mud house with a thatched roof, located by a riverbank a few kilometers from the park's management building. They got married, had two children, and gave their all to protect the rhinos and other animals at Garamba.

Their two dogs accompanied them on their move. Hers was soon eaten by a crocodile, while Fraser's died from a twisted stomach.

It is no exaggeration to say the two of them started from scratch. There were no roads at the park, so Fraser offered to make some. The pair of them would drive their jeep to forge a strip through grass taller than the vehicle's roof. Fraser built bridges and patrolling stations, maintained radio connections, took care of the rangers' equipment and vehicle maintenance and more. He still found time to helm the park's anti-poaching unit, along with the park's manager. Kes went out in the scorching heat to collect data on the local northern white rhinos. She set up and trained the research and monitoring unit, wrote countless reports, and raised funds.

However, to function in such a perilous and politically complex environment, they needed backing from international conservation organizations.

For several years, Kes helmed the already-mentioned African Rhino Specialist Group. She also served as the coordinator of UNESCO's Garamba program. She closely collaborated with Institut Congolais pour la Conservation de la Nature, the Congolese national conservation organization. Along with her uncompromising warrior spirit, her international contacts helped her stare down many a blood-thirsty warlord.

When Kes first saw northern white rhinos in the wild, they reminded her of a mixture between the southern white rhino and the black rhino. "They keep their heads a little straighter, they've got smaller nostrils and also smaller horns." She could now easily list the key differences.

The savannah at Garamba was made of tall grasses, sparsely strewn brushland, trees, and numerous rivers—which made it the ideal home for elephants and northern white rhinos. The lower grasses provided nourishment, while the taller ones could be relied on for cover while resting.

The Garamba National Park covered 4,900 square kilometers of land. Following the Hillman-Smiths' arrival, the number of northern white rhinos residing there doubled to thirty or so.

"It was such a joy to watch them reproduce and take care of their young ones," Kes recalled fondly. "I got to meet and befriend every single northern white rhino living in the wild. They formed a clannishly structured society. We took the trouble to map out each of their family trees as far as six generations back. I'm proud to say I was an important part of their family lives."

As she spoke of those years, the red-headed biologist kept smiling like a wonderstruck child.

Back then, saving the northern white rhino also meant shaping the wider picture of African wildlife conservation. The northern whites had long been seen as one of the local habitat's most important species. They served as landscape architects of sorts; their feeding and excreting habits contributed a great deal to the biodiversity of the local flora. Their enormous daily hunger for grass also helped prevent natural fires, especially in combination with their habit of biting off the grass stalks so low that fire could not scorch the plant. It is worth mentioning that removing the rhinos from the grasslands eventually caused a 50 percent decrease in low-grass pastureland.

The introduction of systematic animal monitoring and increased security at Garamba also proved highly beneficial to the other animals, most notably the elephants. From 1976 to 1984, their number dropped from 20,000 to 5,000. By 1995, it bounced back to 11,000. The populations of buffalos and giraffes started to recover as well.[8]

At Garamba, Kes and Fraser were replenishing life. They were writing their own book of genesis.

They were building a new world.

Monitoring the northern whites from the ground and from the air was the most time-consuming of Kes's duties. Nonetheless, she performed it with relish and dedication.

Above all, she learned to love her aerial sorties.

"My favorite time to soar over Garamba was in the early morning, when it was still fresh, and when visibility was at its best. In the mornings, you could see many more animals than at any other time, which made them easier to count. Those were some of my very best memories. Of course, this was in the years before the wars broke out," Kes recalled over an evening glass of gin and tonic.

She was leafing through a stack of photo albums depicting her childhood, her time at Garamba, and her two children. For a while, with war raging all around, the task of schooling them fell to Kes.

"Juggling work with family life was stressful," she smiled. She would always get up early and depart for the monitoring planes. From the front position, she counted the animals herself, while a pair of colleagues kept watch out the plane's sides. They divided the grounds into a grid, with each plot receiving at least one inspection every month.

By 1995, the number of the northern white rhinos at Garamba increased to thirty-two. Yet, this relative idyll did not last long.

At the start of the 1990s, 80,000 refugees sought shelter in Zaire from the civil war in the neighboring Sudan. Many were armed. They settled in the area around Garamba and began hunting the park's animals, using Kalashnikov automatic rifles as well as grenades and rocket launchers. They slowly forged their way towards the south of the national park, where the northern white rhino population was at its densest.

In 1996, Rwanda joined the slaughter in Zaire, aiming to bring fleeing rebel groups—the so-called *génocidaires* responsible for the Rwandan genocide—to justice.

Zaire was completely destabilized by the slaughter in Rwanda, especially in its east. Laurent Kabila, head of the insurgents against the Zairean government, embraced the turmoil by toppling the long-reigning dictator Mobutu Sésé Seko. The country was renamed as the Democratic Republic of Congo. Then sheer chaos broke out.

This marked the beginning of the first Congolese civil war. At the height of the conflict, eleven African countries were involved, along with outside interests much too numerous to list.

Kabila's soldiers entered Garamba in February 1997. They cut a bloody swathe through the park and laid waste to everything in their path.[9] When Kes and Fraser flew over Garamba in June, the ground was strewn with animal carcasses, though the rhinos seemed to have gotten off easier than many other species. They must have hidden in the tall grasses; the Hillman-Smiths managed to count twenty-four survivors— and even five new young ones.

Many of the other populations, however, were absolutely devastated. The hippos were reduced to a quarter of their former number and the buffalos to a third.

Kabila's mercenaries also destroyed Kes and Fraser's home. The couple had been fortunate enough to flee the carnage, taking refuge at the Nagero camp, which was built by Fraser to house the visiting tourists. They also managed to save most of their crucial belongings like computers, photo archives, and stun guns.

During the ensuing cease-fire, Kes and Fraser strived to quickly reassemble their lives, reequip the staff and reinstate the quick-response ranger unit even though, unlike them, the main instructor had opted to run for his life.

In August 1998, the Democratic Republic of Congo was once more swept up in a wave of violence.

This time, it lasted for five years and brought near total destruction. Approximately six million people died—among them President Kabila, who was killed by one of his bodyguards. For a short time, Garamba fell into the hands of the infamous Lord's Resistance Army from Uganda. Forced to leave the country, the Hillman-Smiths retreated to Kenya.

Yet, they still refused to leave the northern white rhinos high and dry.

In collaboration with the managers of five other Congolese national parks and UNESCO world heritage sites, they set up a group tasked with fostering biodiversity in the regions torn apart by armed conflict. The United Nations allocated four million dollars to the group, which was coordinated by Kes. Some three quarters of the amount were used to pay for armed protection of the animals out in the field.

During the second Congolese war, a squad of rangers kept operating within the Garamba National Park. The rhino and elephant populations experienced a slight increase—10 percent per year in the case of northern whites), but the truce in neighboring South Sudan also gave rise to new enemies—like the Janjaweed, the mounted militia from north-western Sudan.

In April 2003, she made a tally from the air and counted thirty northern white rhinos. By next year's July, there were less than ten. During this interval, civil war erupted in Darfur in the west of Sudan, and the Arabian Janjaweed horse-riders played one of the key roles. Among other atrocities, they also made frequent raids on Garamba and murdered many of the park's animals.

The rangers were issued hand grenades and rocket launchers. Some were killed; many were wounded.

Many of the poachers, however, met their demise at Garamba as well. During the worst of it, the park was a free-fire zone for all sides. Shooting back was the only way to survive.

"Oh, how important it was to be able to trust your rocket launcher!" Kes proudly recalled those infinitely troubled times. "Myself, I was of course not allowed to grab a weapon. But still. . . ."

Over the second half of the 1990s, the northern part of the Democratic Republic of Congo was transformed into a giant abattoir. Even during the height of the violent clashes, however, Kes kept up her aerial sorties. She and Fraser saved the lives of both humans and animals, surmounting seemingly impossible logistical obstacles. They were sometimes helped by a pair of Ukrainian pilots and their Antonov freight plane, flying under the aegis of the UN World Food Program.

On countless other occasions, Kes and Fraser were all alone in trying to navigate a safe path between various warlords, paramilitary units, international mercenaries, corrupt local leaders, and common criminals. They somehow always made it through in one piece.

Most of the northern white rhinos were not so fortunate. One after another, they kept dying before Kes's eyes.

The poachers were responsible for a great deal of the killing. A butchered rhino simply meant money in the bank to them. The war also caused a marked shift across the most favored trade routes. Northern white rhino horns were now exported through Kinshasa, Uganda, and both Northern and Southern Sudan—first to Yemen, and then onward to Asia.

Poachers were far from the only ones exterminating the northern white rhino. Many deaths could also be chalked up to a succession of heavily corrupt authorities in Kinshasa.[10]

When the northern white population dwindled to less than ten, it was clear that the remaining ones would also succumb, barring a swift intervention.

John Lukas, president of the International Rhino Foundation—the Garamba park's key financial backer—proposed translocating five of the rhinos to safety. Kes and the other experts from the African Rhino Specialist Group chose the Kenyan Ol Pejeta Conservancy as the best available location. The key factors were its proximity to Garamba, its level of security and the compatibility of its natural habitat. But the Congolese environmental minister prevented the move by spreading false news to the tune of "the foreigners are once more stealing our natural wealth!"

Three years later not a single northern white rhino could be found in the wild. "Unfortunately, we were also unable to gather any genetic material at Garamba," Kes Hillman-Smith told us in Nairobi.

In 2005, the national park was temporarily stripped of all international financial aid. Kes and Fraser were forced to leave their home. The hope that a living northern white rhino might yet be found in the wild kept smoldering for a while. The Hillman-Smiths simply refused to give up on the notion. But they eventually realized there was no use in fooling themselves.

"We had done everything we could. But we were powerless before the corrupt national government," Kes's voice rang out into the dark.

The International Rhino Foundation later switched its Garamba financing to the African Parks Foundation, which still manages the park today. Garamba's seasoned and well-armed rangers have made sure that the park's elephant population is now on the rise. Garamba's management is currently considering bringing in a few southern white rhinos.

Understandably enough, talking to Kes often felt like talking to a war veteran or a traumatized war reporter. At times, it was clear she was trying her best to cover up her psychological wounds.

In more ways than one, Kes actually *is* a war veteran, though she wouldn't put it like that. Most of our questions about her personal feelings were swiftly deflected; Kes would switch to a completely different topic—the more unrelated the better.

Opening up was not something she found easy to do.

We had to ask our questions as delicately as possible. Sometimes we had to ponder and decode every sputter or barely audible word. Kes's tone remained remarkably even while recollecting both the worst tragedies and her fondest memories. She took every available opportunity to shift the subject to everything that barked, roared, keened, and twittered across her wondrous garden.

Her eyes grew brightest when she talked of her two children and the many animals that kept her company at the family estate. During our stay, Soldier and Flea would occasionally try to impersonate a pair of seasoned guard dogs. But their hearts were much too soft for them to frighten anyone. Even the housecat could pretty much do what it wanted with them.

Just as with the Garamba rhinos, Kes had named all the wild boars roaming her estate. From afar, their curved tusks and gray skin made them slightly reminiscent of rhinos.

One of the boar mothers was named Big Sis. She wasted no time in asserting who ruled the thicket. When it got a bit darker, a few Senegal bush babies descended from the trees onto the verandah. These very shy huge-eyed primates happily munched bananas right out of Kes's hand. They had also been given names. Kes could easily identify each one by their characters and physical traits.

The animal world had always been much closer to her heart than the world of humans.

"Kes is a formidable woman," Douglas Adams wrote about her in 1988. "She looks as if she has just walked off the screen of an adventure movie: lean, fit, strikingly beautiful, and usually dressed in old combat gear that's had a number of its buttons shot off."[11]

The author of the cult science-fiction series *The Hitchhiker's Guide to the Galaxy* visited Garamba while he was writing *Last Chance to See*. The creation of this painfully prescient book took him and the biologist Mark Carwardine halfway around the world on their search for critically endangered species. Thirty-five years on, all we can do is nod at their every word.

At Garamba, Adams was able to see eight of the twenty-two still living northern white rhinos. Kes had never heard of him before his visit. Yet the cult author had no trouble winning her friendship through a combination of wit and earnest clumsiness. When he left, she took to reading his books and became a great fan of them.

In the guestroom toilet at her house, the visitors still enjoy the privilege of browsing through Adams' humorous dictionary *The Meaning of*

Liff. A few copies of *Last Chance to See* have even found their place on Kes's bookshelves, where scientific literature reigns supreme.

Back then, thirty-five years ago, the brilliant Englishman refused to cave into the pessimistic projections. He was convinced the northern white rhino could be saved in the same way their cousin, the southern white rhino, had been saved: by moving them among the various national parks.

"The point is that we are not too late to save the northern white rhino from extinction," he wrote in the wonderfully naive *Last Chance to See.*

Sadly, we failed to follow up on his advice. Just like the northern white rhino, Douglas himself departed prematurely for his special table at *The Restaurant at the End of the Universe.* He was only forty-nine.

"It was horrible, leaving everything behind—it was absolutely devastating! Everything was lost, everything got destroyed and all that was left was this huge void. They murdered all the northern white rhinos. You fight and fight and fight and all you've got to show for it is the realization of how powerless you really are," Kes related through clenched teeth.

This was the testimony of the woman who had seen and experienced all the evil humanity is capable of.

"I don't think we really failed at Garamba," she went on. "But we most certainly didn't succeed. We didn't have nearly enough political power. The pain is overwhelming even today. But fighting for something to the very last means loving it with all your heart. And that's the meaning of life, isn't it? At least for me." Her face froze for a few seconds.

Despite all her losses and disappointments, Kes has somehow managed to hold on to her hope, her dignity, and her will to live. Her words and general demeanor betrayed not the slightest hint of cynicism. She had never been one to thump her chest to showcase her heroic role. In fact, it was quite the contrary. To our minds, all this made her the very definition of a heroine. She values her peace and privacy. Yet—some would say miraculously—she never succumbed to hating humanity.

When we first knocked on her door, she had just returned from a dance class.

"Come, I want to show you something!" she beckoned before breakfast, ushering us out for another walk with Soldier and Flea. The evening before, she had mentioned something that sounded like *pyjama* lilies, but we were clueless as to what she actually was talking about.

A huge wild boar was blocking the hole in the wire fence, the unofficial passage from the family estate to the giraffe sanctuary.

"This is the dominant male," Kes cautioned. "So, let's make room for him to pass." Following her example, we started to retreat so the boar could swagger by toward Kes's house, while we slipped through the hole into the preserve.

Very soon, we were off the beaten track, forging our way through clumps of dry grass. We eventually reached what our hostess wanted us to see.

The white-blossomed and violet-striped lily looked as if someone had transplanted it there from one of the world's great private gardens. This was a specimen of the *Crinum macowanii*. We were standing at one of the flower's precious few natural growing sites. It was regularly visited and fastidiously protected by Kes.

"Why is it called pyjama lily?" we wanted to know.

Kes gave us a wide grin. "Because the flowers have the same pattern as the pajamas at the English hospitals."

She took a deep breath of the fresh morning air. Her face was flushed with peaceful gratitude. "Isn't it lovely here? Contact with nature means everything to me. I can't imagine what would happen to me without it."

Suddenly, a huge African harrier-hawk swooped over us and pounced at a nest of small birds at the top of a nearby branch. The bird flapped its wings furiously as it strived to get at the fledglings and the eggs. But then Soldier and Flea arrived at the scene, barking savagely to drive the predator away. Kes also started yelling. The joint spontaneous rescue mission proved a success. The hawk flew off without any discernible reward for its efforts.

Kes looked so relieved that she might as well have been a mother defending her young.

"I dream about Garamba all the time. The park was the center of my world," the legendary conservationist related as we resumed our walk.

She would very much like to write a personal account of her journey, she added quietly. Yet what she'd managed to get down on paper contained very few of her personal thoughts and feelings. We found it difficult to imagine this was ever likely to change. Kes Hillman-Smith wrote like she had always lived her life. Her focus was entirely on those around her.

"Ol Pejeta is a place very close to my heart. I still haven't lost all hope, you know. I honestly believe the BioRescue consortium could save the northern white rhinos. The project embodies what's left of our hopes."

The English-Kenyan explorer spoke the next few sentences with great conviction. "Our species caused the functional extinction of the northern white rhinos. Which makes it our responsibility to try to save them, using all the tools offered by modern technology. It's so exciting for me to observe the progress being made in the assisted-reproduction field. Still, it will be nothing short of a miracle if the program results in newborn rhinos!"

She hesitated for a moment and then pressed on. "Though even this triumph would be something of a half measure in itself, genetic purity hardly teaches the young animal how to live in its natural habitat. An eventual comprehensive solution would probably entail returning the northern white rhinos to Garamba."

Throughout her career, Kes placed great emphasis on the need to treat the rhino population living in captivity and the population still found in the wild as two sides of the same coin.

She tried to persuade the Ol Pejeta team to follow her recommendations on handling the four northern white rhinos that were moved in from the Czech Republic. She wanted to see them treated based on what she had learned in the wild. She remains firmly convinced this could have significantly raised the chances of their producing the next generation. But in her view, the Ol Pejeta caretakers were too protective of their four wards.

Kes's suggestion was to make their lives as harmonious with the social norms of northern white rhinos in the wild as possible.

"In their natural habitat, the most stationary among them were the territorial males," she recalled. "At Garamba, they were at least ten years old and their priority was to defend their territory from the other males striving to mate with the females. The females were moving through the males' territories. When they were ready to mate, the male whose territory they currently inhabited tried its best to keep them there. I'm sure you can guess why. The most mobile were the young males, prowling the edges of the dominant males' territories, until they got strong and brave enough to start challenging for territories of their own."

At Ol Pejeta, however, the new arrivals were treated much the same as they had been at the Czech zoo. The caretakers would release either Sudan or Suni into a small pen and then they would bring in Najin and Fatu.

Kes shook her head in mournful recollection. "I watched Sudan—the older male—enter the pen. He headed straight for the bushes and sprayed them with his urine. Then he voided his bowels and scraped his hind legs on the ground a few times, which were clear signs of territorial behavior. With all that accomplished, he was able to relax and head off to graze and rest. They returned him to the *boma*, and then the two females were released into the pen. They were immediately off to the bushes marked by Sudan. Najin voided her bowels and then she and Fatu—convinced Sudan was near—went to graze and take a mud bath before they were called back to the *boma*. When it was time for Suni, the younger male, to take his turn, he headed straight for the same bushes as well, but only to return without having marked any territory. He looked very nervous. He was clearly still very submissive toward Sudan, even though he was already over ten years old."

Sadly, the described regime proved not at all conducive to successful mating.

Kes is convinced that to help the rhinos get there, the males should have been allowed to create their own territories within two separate pens. Under this scenario, Suni could have gradually started gaining the confidence needed to get him over the hump. The two females would of course have to be moved from one male's domain to the other.

That was what Kes kept strongly recommending, but the Ol Pejeta management would not listen. In our hostess's view, the Ol Pejeta caretakers were also overly protective of the rhinos, often preventing contact between them for fear they might harm each other. This was especially the case when southern white rhinos were present, even though the rhinos' horns had been removed.

"We're talking about a number of missed opportunities here, especially when it came to Suni," Kes concluded. "Though it's also fair to say that the living conditions at Ol Pejeta were generally better than at the Dvůr Králové zoo, given the Kenyan natural environment."

When questioned whether Ol Pejeta could serve as a model for rhino conservation, she took a few moments to reflect.

"Well, a lot of space and money is needed to protect the rhinos," she eventually replied, "which also means a lot of private investment. Among the various types of conservancies, the Ol Pejeta model proved the best one."

Over the last few years, Kes Hillman-Smith began collaborating with the Amboseli National Park in Tanzania, where she immersed herself in mitigating the rapidly growing conflict between the world of humans and the world of wild animals (human-wildlife conflict). When describing her collaboration with the local communities living either inside the park or along its borders, she often used the word "complicated."

"All too often, the conflict between human beings and animals really comes down to conflicts between human beings. Most of the problems inherent in the so-called community conservancy model are related to the 'community' component. You must approach the problems in the way the local communities want them approached. So, you have to make lots of compromises," Kes summed up the situation. "Nothing is easy when it comes to wildlife conservation. There are no shortcuts."

"Is the ongoing campaign at Ol Pejeta, where you visit three or four times each year, what keeps you afloat?" we asked her. We were more than slightly embarrassed to pose such a question, yet Kes seemed to expect it.

"It *is* very important to have an inner driving force," she nodded, staring off into the thicket, "and mine is still firmly anchored to Garamba and the northern white rhinos."

CHAPTER TWO

An Evening at the Rhino Cemetery

THE CEMETERY WAS LOCATED BY A LONE DESERT DATE TREE (*BALANITES Aegyptiaca*) near a macadam crossroad inside the Ol Pejeta Conservancy. We were standing at the burial place of all the Ol Pejeta rhinos deceased over the past nineteen years.

In the late afternoon hours, when the sun had already begun its slow descent over the horizon, we quietly read the inscriptions on the tombstones. It felt like reading forensic reports on some of the most savage murders imaginable. This makeshift chronology of the extermination of the planet's second largest land-based mammal was yet another undeniable proof of human cruelty, a quick course in the collateral damage wrought by our so-called civilization.

The tombstone marking the final resting place of a female black rhino named Ishirini read: "Born May 17, 1996, died February 22, 2016. Probably killed by a poisoned dart. The security team found her in great pain, with horns already removed. She was twelve months pregnant."

Ishirini is the Swahili word for "twenty." The age she reached before she was butchered.

The neighboring tombstone said: "Kaka, black rhino (1996–2015). His body was found the day after he was killed with poisoned darts and had both horns cut off." Kaka is the Swahili word for "brother."

Next to Brother lay the female black rhino Mwanzo, which means "Beginning." Her tombstone simply said: "Mwanzo, 2007–2011. Shot. Both of her horns were removed."

The Ol Pejeta Conservancy is probably the continent's safest place for rhinos. But still, of the sixteen rhinos who perished here after 2004, only three cases could be chalked up to natural causes. Those three were Sudan, Suni, and Morani, the last of which was named for a Maasai warrior.

At the cemetery, Sudan and Suni's graves were awarded an especially prominent location. They looked like twin portals into the recent past of a functionally extinct species.

Today, only the last two female northern rhinos still live at the Ol Pejeta Conservancy: Najin and Fatu. They are better protected than the UK's monarch and the heir to the Saudi throne combined.

At least to us, it seemed the Conservancy's royal treatment of the last two might be a form of penance for all the harm our kind has brought to the rhino.[1]

The Laikipia County's privately owned Ol Pejeta Conservancy covers 364 square kilometers. Stretching along the base of Mount Kenya, Africa's second largest mountain, the Conservancy is located almost 2,000 meters above sea level. It also crosses the equator, so the climate is mild.

The region never gets too hot or too cold. It is also drier than what most Kenyans are used to. The Conservancy serves as both a breeding ranch and a luxury refugee camp for over a hundred different species of mammals, many of them endangered.

The most endangered by far, of course, are Najin and Fatu.

"The project of saving the northern white rhino has always been and will always be costly," we were told by Ol Pejeta's former CEO Richard Vigne. "Many believe our goal is not worth it. They're usually so quick to dismiss our efforts! 'You're saving one species, while thousands are lost!' they say. But we strongly disagree. And we are very determined to follow the project through."

We met up with Vigne at the Conservancy's seat in April 2021, just as a brief shower was announcing the arrival of a belated rainy season. Vigne's tenure as Ol Pejeta's CEO lasted for sixteen years. It ended in 2021, when he took over the helm of African Leadership University in Kigali, the first Rwanda-based educational establishment for Africa's conservation leaders.

According to our host, the northern white rhino project had already brought a lot of much-needed attention to the animal extinction problem. Global awareness was certainly on the rise. Just the public attention drawn to Sudan—the most eligible northern white rhino bachelor of his generation—had brought in millions of dollars in donations, as estimated by Ol Pejeta's public relations partner.

"Sudan is now a global icon!" Vigne enthused. "This was a great victory for the conservation cause!"

Over the last twelve years, Ol Pejeta spent around two million dollars on the northern white rhino conservation project. The plan is to spend another ten million over the coming decade.

"Is that really so much?" the Kenyan entrepreneur and conservationist demanded of us.

"Ten million is the yearly salary of a mediocre defender in the English Premier League," we replied.

"Precisely," Vigne nodded. "We are, after all, talking priorities here!"

It is worth noting that in the global context, rhino conservation projects are generally awarded incomparably less funding than projects focusing on saving lions, elephants, and other large land-based mammals.

During British colonial rule, the Ol Pejeta grounds were used as a livestock ranch. Its entire savannah landscape was divided into enclosed grazing pens. All wild animals that encroached on the territory were immediately killed.

Back then, this was only par for the course.

Ol Pejeta's task was to produce food for the still fairly young British colony. In those days, it was believed that the coexistence of ranchers and wild animals was impossible. This belief was a self-fulfilling prophecy. The wild animals were seen as competition, so when building a ranch, the builders simply slaughtered everything within a large radius.

In any case, a lot of new infrastructure was built across twenty square kilometers of private property. Each area boasted its own water source. The residing livestock was penned in, but were allowed to graze without direct supervision.

"Those ranches were extremely profitable," Richard Vigne noted.

By 1977, wild animals were therefore almost completely gone from Ol Pejeta's territory. However, after a ban on hunting was put into force, their number started to grow. Especially the elephant population, which used to be critically endangered, but started quickly recovering after Kenya banned elephant hunting in 1973.

The elephants would normally merely cross the Laikipia County, yet all the wars and instability in the north and the west started to make them linger in the province, like refugees.

"This meant trouble for the ranch owners. The entire ranching system was forced to adapt. In the meantime, the pastoralists—the traditional shepherds—started returning to the area. This type of livestock breeding was neither efficient nor profitable. By the end of the 1980s, the ranches were barely making ends meet. And the number of wild animals was still growing," Vigne explained the history that led to the founding of Ol Pejeta.

All the changes in the region put the ranchers in dire straits. Searching for additional sources of income, they decided to try their hand at conservation-themed tourism.

In 1988, the ninety-seven square kilometers of ranchland at what is now Ol Pejeta became the core of the Sweetwaters Game Reserve, where several black rhinos were settled. By that time, Kenya's black rhino population had fallen under 300; the poachers were still dropping them left and right. The state-run Kenya Wildlife Service therefore decided that all of the black rhinos had to be moved to the reserves. The heavily protected Sweetwaters easily met all of the requirements.

The black rhino inhabits a diverse range of habitats in sub-Saharan Africa—from Namibia's desert-like savannahs to humid wooded areas. Between 1960 and 1995, poaching reduced their number by almost 100 percent, to less than 2,500 animals. But with the help of well-planned conservation programs, the black rhino population slowly grew to 5,600.[2]

By 2022, the Ol Pejeta's starting population of twenty black rhinos had grown to 149. The Conservancy now serves as the largest sanctuary for black rhinos in Eastern Africa. The populations of the numerous other species sheltered here has recovered as well.

In 2003, the ranch and its neighboring territories were purchased by Flora and Fauna International, a British environmental NGO, with financial help from the Arcus foundation.

The grounds were transformed into a reserve, though livestock breeding remained a part of the business model. In 2005, the Conservancy's long-term management was transferred to a Kenyan non-profit organization; Richard Vigne took over as CEO. The thriving business model he helped set up comprised livestock farming, environmentalism, and tourism. It is now being imitated elsewhere in Kenya.

"They told me it was a dumb idea and that we were sure to fail," Vigne recalled with a winsome smirk. "To hear them tell it, no tourist wanted to stare at cows at a wildlife reserve. Well, the last fifteen years sure proved them wrong. We've managed to harmonize wildlife, livestock, and tourism. Indeed, we seem to have hit on a brand-new business model! We employ about a thousand people, we pay a boatload of taxes, and we're helping to maintain the ecological equilibrium. For the local community, this is pure win-win."

The Kenyan-born Richard Vigne hails from a British colonial family with a tea plantation to its name. He first got a degree in biology, and then a master's degree in agricultural management in London. Before arriving at Ol Pejeta, he used to run a safari management firm in Uganda and the Democratic Republic of Congo: the home of the northern white rhinos.

In Vigne's words, Ol Pejeta managed to avoid the bane of numerous conservation enclaves in Kenya: namely the conversion of the wilds into farming land.

In the years before World War II, Kenya was populated by five million people. Today, eight times as many live there. The country's dramatic population rise resulted in a growing demand for food. This trend is unlikely to change for years to come.

The expansion of farming areas, however, proved very inimical to the environment and its biodiversity. Wildlife bore the brunt of the transformation. After all, the peaceful co-existence of the two worlds is only possible through exceptionally sound management.

The Conservancy has long become a vital part of the Laikipia County ecosystem. Along with the Maasai Mara national game reserve, it boasts the highest wildlife density of anywhere in Kenya. Among other species, there is a growing population of southern white rhinos—forty-one of them—at the Conservancy. The southern white rhino rates as potentially endangered on the IUCN Red List.

Ol Pejeta also boasts some of the densest lion and hyena populations in the country. It hosts Kenya's one and only chimpanzee sanctuary.

What can be seen and experienced at Ol Pejeta is a hardly untouched Africa. It is a model of adapting to certain exigencies, especially steep population rise and the consequences of climate change.

"We need a few business models that can protect the conserved areas and also help the locals survive. Our model is already emulated across the entire Laikipia, in Maasai Mara, in the western Tsavo region. . . . Of course, in some places it proved more successful than in others. Managing private property like Ol Pejeta is generally much easier. But we proved that the model works," explained the Kenyan businessman and conservationist, who wasted no opportunity to stress the importance of close collaboration between wildlife conservancies and local communities.

The first years of the Conservancy's operation saw several disputes with the locals. Relations remain far from ideal to this day. "I don't think they will ever fully accept us," Vigne shrugged. "On the other hand, we provide employment to about a thousand people, and we contribute a lot of taxes to the local budget. In general, I think our relations with the local community are very sound."

In these parts, nature conservation has traditionally been linked to philanthropy, but Ol Pejeta's management has successfully framed it as the driving force of sustainable development.

"This is only possible if the profits of our work are also spread beyond the conservancy's borders," Vigne noted. "In order for us to function well, the support of the local community is paramount. We are backing thirty different schools. Our scholarships are received by two hundred students. We're helping the local hospitals. We're building roads. We are doing everything we can to tackle the ever-mounting conflict between the wildlife and the world of humans. You see, so much still needs to be done!"

Vigne claims that Ol Pejeta really serves a single purpose—that of nature conservation, with a heavy focus on preserving biodiversity. Of course, the management also needs to make some concessions to the fact that they are running a business.

"We have a number of tools at our disposal," Vigne explained. "The fact that we've become a globally recognizable brand is a great help. To remain true to our goals, we had to work very hard to overcome our shortcomings. We never stopped innovating. We still draw a lot of criticism from the more traditional conservationists, but much less than we used to! You must realize that in those circles, the very word 'profit' used to be taboo. We see the merging of natural conservation with sound business practice as a winning combination—as long as the profits are invested in protecting the wildlife."

Vigne cleared his throat. "The office you're sitting in right now saw countless conceptual disagreements," he revealed. "Our project focusing on Najin and Fatu actually gave rise to accusations of technological wizardry. Or let me give you another example. When Sudan died of old age, I happened to be in London, so they said I was there on purpose, to milk the death of the last northern white rhino male for publicity. But let me tell you, we know very well that all of us—even the large business companies—depend on the successful protection of the natural environment!"

The Ol Pejeta Conservancy is protected by 120 kilometers of high electrical fence. Najin, Fatu, and their southern white peer, Tauwo, are afforded even more safety inside their pen, which is about three square kilometers wide and constantly monitored from the outside by armed guards. Within these confines, the rhinos' well-being is ensured by their designated caretakers.

On average, protecting one rhino costs around 10,000 dollars per year, we were told by Richard Vigne. Now, 191 rhinos reside at Ol Pejeta. "From the outset, we understood their maintenance would be very costly—regardless of how well tourism was doing," Vigne smiled.

In the past, the Conservancy was sent rhinos from various other reserves and national parks in Kenya. "It wasn't only about the money. All the poaching was truly devastating for the rhino population. Especially in the 2007–2008 and 2012–2014 periods. During those times, a number of smaller reserves decided to send their rhinos here, given how vulnerable they were in their former locations. It was very dangerous. The risks to our staff were enormous as well," Vigne recalled.

In 2020, the Ol Pejeta Conservancy quickly came close to having the highest sustainable number of rhinos in residence. According to Vigne, to increase it any further would be detrimental both to the rhinos and to the Conservancy itself.

Yet there are no plans to relocate any of the current animals. The harsh truth is that there is no room for them anywhere else in Kenya. The Conservancy recently purchased some eighty square kilometers of adjacent land from the local community. The management wasted no time in protecting and fortifying the newly acquired grounds, in order to start settling them with potential surplus animals.

"We're hoping to move a part of our rhinos there very soon," Vigne remarked on the Conservancy's future plans. "The whole thing will cost at least $3.5 million. But it will provide living space for an additional forty rhinos. At least that many will be born here over the next three years. Whatever happens, the new land should come in very handy."

The project of saving the northern white rhino might soon make Ol Pejeta even more recognizable than it already is. In this context, Vigne already saw the project as a great success, since Najin, Fatu, and the late

Sudan had already brought in substantial funds and recognition to the conservation cause.

"That was a huge achievement! As for the scientists, well, they're bound to succeed sooner or later. The real problem is the extent to which an endangered species can be preserved. It takes several decades to create a vital population. Only then will we be able to say the species had been saved. I don't believe this is likely to happen within our lifetimes," Richard Vigne coolly and rationally laid out his views.

The Covid-19 pandemic took a heavy toll on the Conservancy's income; it was derived of tourism. One hundred and fifteen thousand people came to visit in 2019, while the following year only saw 30,000. Fifty-six percent of those were locals, who understandably spent much less than visitors from abroad.

The projection for 2020 was that tourism would bring in around $12 million. The actual number was 1.6 million.[3] 2021 also ended in a huge loss. However, in February 2022, when we visited Ol Pejeta for the second time, it looked like tourism was starting to thrive again. The Conservancy was slowly digging itself out of the worst of its financial hardships.

Over the difficult period marked by the pandemic, it was ranching that pulled them through.

"Both years, it ended bringing in almost a million in profits," we were informed by Richard van Aardt, head of the Ol Pejeta Livestock Program, as we joined him on a brief tour of the Conservancy's pastures. At any given time, about 7,500 cows and calves of the indigenous *boran* breed are grazing on the reserve grounds. Their meat, dubbed "conservation beef," is sold at a premium to butcher shops, hotels, and elite Nairobi restaurants.

Lion attacks cost the Conservancy about 100 head of cattle per year, though cows are not the kind of animals to simply surrender to predators. After all, they've managed to coexist for 2,000 years.

"When they smell lions, they press together. If they're attacked, they disperse. Each herd also has two or three shepherds to protect it," explained van Aardt, who was dressed like a cliché of cowboy lore. He,

too, saw the coexistence model devised at Ol Pejeta as a winning combination, satisfying both the growing Kenyan demand for food and the
wild animals' need for protection.

As if to second the view, a black rhino female with a young one in
tow appeared roughly 500 meters from the mooing cow herd.

Many of Kenya's rangers and conservationists believe that the consequences of climate change fueling the conflict between the world of
humans and the world of wild animals will now pose a greater threat to
the animals' wellbeing than the poachers. This view seemed especially
pertinent at the beginning of 2022, when virtually all of north of Kenya
was ravaged by drought. It was the worst one in the last forty years.

According to UN data, as many as half a million undernourished
children were living in the Kenyan north at the beginning of 2022. The
region was regularly beset by droughts before, but the really bad ones
used to come in five-to-ten-year intervals. These days, radical drought
seems to be on the cards every other year.

Much the same could be said of all of East Africa—Kenya, Tanzania, Uganda, Burundi, and Rwanda—where, in January 2022 some
twenty-six million people faced starvation, according to the UN's World
Food Programme. A total of 2.4 million of them lived in northern parts
of Kenya.[3]

In countless places all over the country, not a single raindrop has
touched the ground for a year. The rainy season already was regularly far
overdue, but this was the first time it failed to appear altogether.

The scant water sources completely dried up. In 2021, the south of
Kenya was devastated by floods, while locusts destroyed most of the 2020
harvest. The wild animals' migration paths from the north to the south
were rapidly being severed. The eminent Maasai Mara national reserve
saw a great downturn in local animal migration. Even the wildebeests'
mass migrations, considered by some as the greatest spectacle in the
entire animal kingdom, were greatly diminished. In 2021, this happened
for the first time in recorded history, we were told by local rangers and
conservationists.

One of the few species whose migrations did not take a downturn was our own. According to the latest World Bank data, the consequences of climate change will drive some forty million East Africans from their homes by 2050.[4]

Kenya is and will be one of the countries hit hardest by climate change. Projections show that, between 2000 and 2050, its average temperature should experience a 2.5°C increase. Over the last two decades, Kenya's temperature already rose by almost 1°C.

It is important to realize Africa hosts 17 percent of the world's population, while the continent's combined CO_2 output into the atmosphere amounts to a mere 4 percent of the world's total.[5]

Over the last few years, armed conflicts have broken out across areas where groups of traditional farmers and shepherds have been fighting for water for centuries. The worst of the violence took place at the Kenyan border with Uganda and South Sudan. In 2021 alone, these conflicts between various ethnic groups killed hundreds of people.[6]

For several years, the consequences of climate change were also the driving force behind the conflicts in the nearby Horn of Africa—namely in Somalia and Ethiopia. The latter country's initiative to build a dam on the Blue Nile opened the door to wars with Egypt and Sudan. People and animals were slaughtered left and right. In Kenya's northwestern Wajir County, 100 giraffes were lost to starvation by the end of 2021. Some of them died in horrible pain, being too feeble to drag themselves out of the mud.[7]

The image of their demise recorded by the photographer Ed Ram can be interpreted as a savagely accurate forecast of what lies in store: a shape of things to come. It can also be viewed as a paraphrase of the infamous apocalyptic image of a starving boy (long mistaken for a girl) and a vulture taken in 1993 by the Canadian photographer Kevin Carter in South Sudan. The photo won Carter the Pulitzer Prize, which didn't prevent him from committing suicide shortly thereafter.[8]

Over a period of two years, the brutal drought turned pasturelands into desert and forced some Kenyan shepherd groups to move south in search of greener and damper areas. At the beginning of 2022, several hundred shepherds and their cattle left Samburu County to travel the approximately 200 kilometers to Ol Pejeta and other Laikipia private reserves, where water was traditionally much less scarce.

Yet even this can no longer be relied upon. The Mount Kenya glaciers, the main source of water in this part of Kenya, along with snow, are rapidly melting. Meteorologists predict that the eleven glaciers on the magnificent mountain above Ol Pejeta will melt away in thirty years at the latest.[9] Even now, hardly enough of the legendary snows of Kilimanjaro are left to provide a proper sample.

This is nothing less than a declaration of war. A war on both humans and animals.

By the end of 2021, the drought was so severe that several Laikipia reserves like Ol Pejeta decided to provide shelter behind their tall fences for a certain number of shepherds and their cattle. But relations between the private reserves, and the locals, who are wholly reliant on nature, failed to improve, at least to a sufficient extent. The private reserves are able to extract their own water and can overall provide incomparably better living conditions for both humans and animals.

Forever inflaming the never truly healed colonial wounds, these conflicts are only mounting and assuming the dimensions of a class war. The deteriorating climate conditions are a crisis long-turned permanent. This permanent crisis has already done a great deal to reinforce Africa's societal divisions.

In places, the shepherds have already cut through the fences and entered the reserves, convinced that the managers had appropriated land which used to belong to the community. By the end of 2021, around 6,000 cattle and 3,500 goats and sheep were grazing without the management's permission within Ol Pejeta's Mutara district alone.

"People are coming up with the notion of conservation as a new colonialism," viewers of the *CBC* network were informed by Yoakim Kuraru, head of a shepherding community surviving on the edges of Ol Pejeta.[10]

"Wild animals are valued more than us. We think that land has been fenced off so that we, we the pastoralists, cannot access that [private] land," Kuraru added.

His claims seem hard to refute, especially in the context of historical grazing rights. It is also getting increasingly hard to overlook the conflict's racial component, even more salient given the fact these are the more prosperous parts of Kenya.

In the region, the old cliché stating that most reserves are run by whites while most poachers are dark-skinned is a veritable axiom, powered by the need to survive. It is also true that despite their noble conservationist role, many of the reserves are functioning as ivory towers, and some even like occupational outposts.

"Water is everything," we were told by Jackson Lenamaita, head of a local community residing on the outskirts of Ol Pejeta.

It was February 2022, and we were standing in the scorching sun next to a wildlife corridor used by the animals for their daily migrations between the reserve and the outer world. During our visit, the migrations only seemed to flow one way: from north to south, from hell towards a sanctuary with enough food and water still left.

The elephants came as well. They had trudged here all the way from Samburu, their origin discernible by the characteristic red soil clinging to their skin.

"Life used to be predictable. Just like the weather and the climate. We knew when it would rain, and when the drought was due. We could make plans. For the past several years, it hasn't been like that. There's no predicting anything. Life has grown horrendously hard," continued the local tribal leader, who'd served as an UNPROFOR soldier in Croatia and Bosnia and Herzegovina in 1995 and 1996.

His age of fifty-seven—he looks much older—makes Lenamaita one of the oldest members of his community. Astoundingly, the average age of the Kenyan population is seventeen. When it comes to the countries with the youngest populations, eight of the top ten are located on the African continent.

"We're fighting to the best of our abilities," Jackson grimaced. "But if I'm honest, I can't begin to imagine how we'll be able to survive. Every year it gets worse. The number of people is growing, and the living space is contracting. This goes both for the people and for the animals. The conflict keeps getting worse, along with the consequences of climate change. What we really need is water. Tell the world we need water!"

For Jackson, broaching the subject of there being too many people was like kicking a hornet's nest. If he were to merely mention his belief it was high time to start discussing birth control at one of the tribal elders' meetings, he would be "torn by human lions," he revealed. The subject remains an absolute taboo.

"The only solution to that particular problem is education. But what can we do, when so many families are unable to afford sending their children to school? It is too expensive. And our community grows poorer every year," Lenamaita explained tiredly.

The community Lenamaita presides over is a thousand strong. It is entirely dependent on a single well. Even this single source of water was only recently provided by Ol Pejeta's management as an exercise in buying social peace.

The hottest part of the day arrived, and a group of children gathered around the well. Some were as young as five or six. They had brought huge plastic canisters, which they were now filling from a leaking pipe.

"This is our one source of water!" Lenamaita lamented, looking out at a landscape that was terrifyingly dry on all sides—well, on all sides save one. We were standing by the tall, electrified fence separating the local community from the much greener Conservancy.

"If an elephant at the reserve were to step on one of the pipes, we'll be left without water," Lenamaita pointed out.

To make matters worse, the traditional shepherding community is also heavily reliant on selling their livestock. Over the last few years, however, the prices plummeted. A cow used to bring in €700. Today, the price is around €250. The price of sheep fell from €300 to €50 The scarcity of food and water has turned the animals paltry and ill. On the day of our visit, the only things being sold at the Mutara marketplace were used clothes and a smattering of locally produced fruit and potatoes. Food prices have long gone through the roof.

Everywhere we went, a single word kept springing to mind: *Conflict. Conflict. Conflict.*

Just a few days later, war broke out in Ukraine, one of Africa's two greatest suppliers of wheat. The other one is Russia.

Nighttime was falling over the savannah. Ol Pejeta's buffalos, zebras, gazelles, and impalas were clustering in groups, seeking safety in numbers for the night. A large group of water buffalo had gathered at the lake to tank up on liquid.

While driving around the compound, we spotted the youngest of the southern white rhinos. Born a month before our visit, the still-hornless gray fellow seemed at the height of playfulness. He kept kicking and ramming into his parents, who were grazing stoically in the dusk, until the mother finally had enough and pushed him away with her horn, sending the baby flying at least a meter to the side.

It was to no avail. After he picked itself up, the young one's evening frolic went on undeterred.

When we were nearing our lodgings for the night, a magnificent lioness crossed our path in the dark blue dusk. Adopting a slow graceful lope, she kept calling out yearningly to her mate. Only a few meters off, the thicket, briefly illuminated by our headlights, revealed the courting ritual of another lion couple. When the lioness took in the scene, she turned and ran off into the night. For a moment there, we could swear she looked disappointed.

NAJIN AND FATU'S CARETAKER

"We're sorry you had to get up so early on our account," we apologized to Najin and Fatu's caretaker Zachary Mutai. Meeting him in front of the pen where the two rhinos spend their nights, we were still hazy and sleepy-eyed from getting up before the crack of dawn.

"No problem!" Zachary beamed, looking as fresh and rested as the morning taking shape around us. "I normally get up at five. At six I'm already with the rhinos. They're waiting to be let out to graze."

The man in typical green ranger gear unlocked the door of the heavily protected pen. After he removed a pair of additional electrical wires, we were free to enter the domicile of the two horned giants, each of them weighing more than two tons.

At our arrival, their heads snapped out of their morning reverie. One of them revealed a long and proud horn, whereas the other one's horn looked more like a decomposing stump. It was an awe-inspiring sight and more than a little bit terrifying.

Along with photographer Matjaž, we instinctively retreated behind Zachary's slight and spry frame.

The first to approach was Najin, the older one. "Najin, good girl," Zachary greeted her in a relaxed and kindly manner, placing one of his palms in front of her nostrils. He encouraged us to do the same.

Rhinos can smell and hear exceptionally well, whereas their vision to the left and right of their horns only reaches up to fifty meters. Najin carefully smelled all three of us, gave a sudden loud snort and then moved on. With her head lowered and swaying in the rhythm of her almost inaudible steps, Fatu was the next to approach and inspect our small apprehensive expedition.

Fatu, the daughter, is generally regarded as the somewhat more unpredictable one. But she was remarkably quick to accept us as well. The last to appear before us was Tauwo, the southern white rhino who's kept Najin and Fatu company for several years now. She and Fatu had been friends since their "primary-school days," so they usually stuck together, while Najin, in her late middle age, preferred to keep to herself. Tauwo, born and raised in the wild, was clearly not one to lock horns with.

"Tauwo is protecting her two friends," Zachary warned.

The caretaker's manner with her was much more careful than with Najin and Fatu. When she tried to come too close, he would swiftly bend down and tear off a twig from the first small bush he could find to start swinging it at her. If that didn't work, he would squat and immediately get back up with outstretched arms, making himself appear bigger than before.

At any rate, our greeting sniff-fest proved a resounding success. We had successfully completed the initiation.

Over the following days, Najin and Fatu slowly got used to our smell. When we left, we were convinced their short memories would surely forget us almost as soon as we departed. However, when we returned to Ol Pejeta ten months later, it felt like a mere day or two had passed. Almost immediately after our arrival, even Tauwo accepted us as old acquaintances—or, translated into rhino terms, as entities which didn't bother her.

Our reunion with Zachary Mutai could not have been warmer. From day one, Najin and Fatu's personal angel guardian struck us as the purest of souls, whose every gesture was a tribute to how life really should be lived. Throwing our arms around him, we knew we had found a lifelong friend.

Zachary Mutai is forty-four years old. He is a slight man with a small military green hat and trousers tucked into always-polished black boots and the head of the six caretakers in charge of the thirty-three-year-old mother Najin and her twenty-two-year-old daughter Fatu.

As already mentioned, Zachary also took care of Sudan, the last northern white male. At the age of two, Sudan was captured at the Shambe National Park in the south of Sudan. Then, he was moved to Czechoslovakia in 1975. When he died in 2018, Zachary was by his side. The last time he stroked Sudan behind his ear and rested his head on Sudan's huge horned head, Zachary's sadness almost filled up his lungs and exploded his heart.

After Sudan's demise, only the aging Najin and the somewhat capricious Fatu were left in Zachary's care. Even though his head barely reaches above their backs, his 'girls'—as the two rhinos are called at the Conservancy—obey his every command. Especially Najin. Each morning we spent in their company was its own lesson in how humans and animals can coexist in perfect harmony.

"When visitors are brought into the pen and the two of them are far away, I call out to them, and they respond like dogs. Especially if I'm carrying one of their favorite treats, like carrots. If they try to attack a vehicle, I can also get them to stand down with a simple command—again, as if commanding a pair of dogs," Zachary described the compliant nature of the two rhino girls.

Over the many mornings and evenings we spent in their company, the girls demonstrated their friendly and obedient nature again and again.

"Respect is the most important thing," Zachary nodded. "The wild animals are mostly harmless if you treat them with respect. Only humans kill for no good reason—for fun, for sport, or for money. For sheer stupid greed."

As he led the way around the grazing grounds, Zachary also shared that he had long learned to think half like a human and half like a rhino. His constant and infinitely tender cooing and caressing revealed a profound love for his two wards. He was intimately familiar with all of their habits and whims. After all, he'd spent more time with them than with his family.

"These two animals are a part of my family," Zachary acknowledged. "They can't speak, but we're still able to communicate very well. I tell them how special they are, how much I love them, and I also tell them to behave."

Over the thirteen years he has spent with the northern white rhinos, Zachary learned their language, based on sound and body posture. He understands Najin and Fatu to their very core. They, he is convinced, understand him at least as well as he understands them.

"I don't have very many stories about myself," he related as the morning got warmer, "But when I'm with them, I am able to share in their stories, and that's infinitely precious to me."

In his own home, Zachary is surrounded by girls as well. He is the father of five daughters and a son. His love for his two families and his passion for his work bring meaning to his existence. He is often so happy that he actually daydreams about his own life.

He freely admits that he is not particularly interested in the world beyond the Conservancy's borders. The fate of the rhinos has taught him everything he needs to know about what human beings call society. Or—"civilization."

A lesser man might have snapped under the weight of everything he had seen and heard. Zachary, however, not only survived it with his soul intact, but also managed to hold on to his good cheer. He is simply too at peace with himself not to be cheerful—and too connected to nature, his greatest teacher.

"The animals lead very straightforward lives," he told us. "If you spend a lot of time with them, you can learn to appreciate the little things in life, just like they do. There's so much they could teach us."

A simple, straightforward life. Consistently and without a hint of boastfulness, Zachary practices what he preaches.

He is always dressed in light clothing, even if he must stick his fingers deep into his pockets on cold Ol Pejeta mornings. Heavier clothes, he says, would only weigh him down. He also shuns the use of binoculars, given that he possesses an in-built natural pair. During our first visit, he spotted a rhino from more than a kilometer away. He was also able to discern the rhino's species.

He would never swap a cup of his favorite black Kenyan tea for coffee, while his favorite dish is *ugali*, a maize-based cake that is often served as a side dish to traditional Kenyan meals. Zachary, however, prefers it as the main course.

He normally spends thirty days with his rhino wards, then six days with his family. When we first visited Ol Pejeta in April 2021, his "shift" had already ended, but the pandemic-induced lockdown of Nairobi County had prevented him from taking the twelve-hour bus ride home.

"I'm starting to really miss my family," he told us back then. "And there's also a lot of work waiting for me at the farm." He and his wife grow tea, most of which they sell. They also grow most of the food the household consumes.

When he is back at the farm, Zachary always stays in contact with the other rhino caretakers, who fill him in on how "the girls" are doing.

His wife and children exhibit no signs of jealousy—in fact, quite the opposite. "Before I head home, I always take a few photos of the Conservancy animals, so I can show them to my family. You know, they're so happy because they know that, through me, they can also play a part in protecting wildlife!"

He is also a great expert on the local birds. As we trailed the pair of enormous horned lawnmowers, who occasionally let out a wet fart, Zachary would often halt, lift a dramatic finger, and hold forth on the nature of the birds we encountered.

Even when he is not escorting reporters, Zachary spends at least an hour each day keeping his wards company while they graze. When he is with them, he never gets bored. All the while, he blissfully drinks in the surrounding sounds of nature.

He is also a great expert on the local birds. As we trailed the pair of enormous horned lawnmowers, who occasionally let out a wet fart, Zachary would often halt, lift a dramatic finger, and hold forth on the nature of the birds we encountered.

We managed to jot down at least some of their names: superb starling, wattled starling, white-headed mousebird. Most of the time, cattle egrets would lurk around Najin and Fatu, opportunistically feeding on the insects swarming the pair. To those cattle egrets, the rhinos must have looked like gigantic mobile fast-food stands.

The red-billed oxpeckers, on the other hand, would frequently catch rides on their backs. Until not so long ago, it was believed these birds formed a symbiotic relationship with rhinos. The red-billed oxpeckers were supposed to provide a valuable service by removing ticks from the rhinos' skin. Yet more recent studies have shown that the birds' pecking actually prolongs the healing of wounds. They are, however, not entirely harmful. They also clean out the rhinos' earwax—in quantities we preferred not to inquire about.

Zachary is also highly knowledgeable about the local flora, though he claims this to be one of his "weaker fields." As we walked on, he pointed out the black rhinos' favorite snack: the *Vachellia drepanolobium*. It is commonly known as the acacia tree, or the whistling thorn. Some of the tree's longer thorns are known to grow bulbous spines, which get inhabited by symbiotic ants. When the wind blows through the ants' entry and exit holes, a series of whistling tones is produced.

Still trailing Najin and Fatu while they grazed, Zachary warned us about the roaring of lions in some far-off thicket. He also drew our attention to the gibbering of the chimpanzees residing at the nearby Sweetwaters sanctuary. He pointed out a few ostriches strutting their stuff and sprinting in the savannah. He spotted a moving herd of zebras, which we were only able to discern as a clump of dark dots far off toward the horizon. Last but not least, he pointed out a well-camouflaged giraffe in the bushes, hardly a frequent visitor to the domicile of the last two northern white rhinos.

With every step and turn, Zachary proved his worth as a respectful reader and interpreter of nature. Even if a large part of humanity seems to have lost their reading glasses; even if most of the rest of us are on the brink of collective blindness.

The leader of the Ol Pejeta caretakers always wanted to become a conservationist. Zachary grew up on a farm north of the Maasai Mara national reserve, the area with the highest wild-animal density in the world. His beginnings were the perfect starting point for what his life became afterwards.

"My father worked at Ol Pejeta while it was still a livestock ranch," Zachary recalled. "I was able to visit him regularly and observe all the elephants, giraffes, and zebras. In 1988, the ranch started expanding into wildlife conservation, so I jumped at the opportunity to get a job. At first, I oversaw the electrical fence maintenance, but I slowly rose to the position of head caretaker of the white rhinos. And my priority has always been to help make these animals feel safe."

Even though he has been serving in this role for more than eleven years, Zachary's priorities remain unchanged. "Safety is still my number one concern."

One bright morning, he held up a tenuous fiber he'd picked off Najin's smaller horn. "Can you imagine?" he demanded. "They're hunting them for their horns, which are nothing else than the keratin our nails and hairs are made of!"

Visibly angry, Zachary paused to regain his composure. "When we shorten them, the horns grow back," he explained through clenched teeth. "But the poachers simply chop them off, along with a huge chunk of the rhino's face. Very few animals survive the process and those who do, of course, suffer permanent consequences."

Up close, Fatu and Najin seemed unperturbed by the wider implications of their species' imminent demise. After observing them every morning right after sunrise, and every evening before the sun vanished behind the magnificent horizon, we concluded they were a pair of happy slaves to routine.

After waking up they linger briefly in their inner pen, where they feast on the freshly sprung grasses. Then they head off for the larger enclosure, always in the same formation. In the mornings, the three-headed column is led by Najin, while in the evenings Fatu takes over. Before entering the larger pen, Najin always glances to the left, where her heartthrob, named Owuan, is penned.

On occasion, the three female rhinos playfully charge each other. But when grazing, they are only interested in fresh clumps of grass. On the pasture grounds, love becomes a one-way street.

When they arrived from the Czech Republic, Najin and Fatu didn't know how to graze. They were also overweight, having been raised on bread, apples, watermelons, spinach, and bananas. Their legs were weak, their muscles atrophied.

It was Tauwo who schooled them in natural behavior, and in a certain ferocity.

When they arrived, Najin and Fatu were unbelievably polite and gentle. The caretakers' job was to teach them to obey commands issued in English and Swahili. In the beginning, they had to talk to them in the Czech language. "*Pojd', pojd' Najno!*" Zachary laughed in merry recollection.

In the mornings, the girls graze from six to around nine. Once it gets too hot, they fill their bellies with water and lie down to rest, only to resume their grazing when it gets a bit cooler, around four or five.

Spells of cool weather bring an additional vivacity to their routine. They seem to adore rain with its impromptu mud baths, which provide the opportunity to remove parasites from their skin—if Zachary hasn't already removed them. The mud also protects them from the sun.

Rhino skin is two and a half centimeters thick, and their horns grow up to 1.2 meters in length. Observing their slow methodical trundle down the pasture, we kept marveling at how two tons of meat, bones, muscles, skin, and horn could be assembled from a purely vegetarian diet. At any rate, Najin and Fatu keep their secret formula to themselves, regardless of how interested the world's vegans might be to hear it.

The murder of a single rhino means a great loss for the entire species, given its late sexual maturity and slow reproductive cycle. In Zachary's opinion, the best approach to preventing poaching is through education. Najin and Fatu's caretakers thus took on an important role in raising both local and global awareness of the importance of rhino conservation. They have long become the voices of the speechless last two.

In the past, Zachary traveled to Durban in the South African Republic, where he presented the project of saving the northern white rhinos to schoolchildren and businessmen. In 2020 *National Geographic* invited him to New York to spread the word. Unfortunately, the Covid-19 pandemic prevented the trip.

The limelight recently cast on the northern white rhino made Zachary a celebrity in his own right—though the term hardly seems appropriate, given his awe-inspiring modesty and humbleness.

In 2018, the year of Sudan's death, *The Guardian* chose him as the person of the year in the field of natural conservation. He was featured in photographs taken for *National Geographic* by Amy Vitale. His winsome smile twinkles from a group photo the northern white rhino caretaker had taken with David Attenborough, who visited Ol Pejeta in 2019 while making his *A Life on Our Planet* documentary. On that occasion, Zachary got to meet his great role model, whom he sort of smuggled into the Conservancy to avoid journalists and rubberneckers.

Zachary readily admits that before meeting the bard of nature documentaries, he suffered from a severe case of stage fright.

He also introduced Najin and Fatu to a number of other celebrities like the Chinese basketball star Yao Ming. Ming promised to spread the word of the project's importance in China, which is one of the countries most responsible for the illegal slaughter of rhinos and several other endangered species.

Even though the last two receive the best veterinary care imaginable, Zachary's eyes and palms are always scanning them for potential ailments.

"You see this rough patch of skin?" he pointed at Najin's hind thigh. "This is where trouble might start. The northern white rhino's skin is very thick, so nothing gets through. But if there are lesions, which lead to inflammation, it can mean mortal danger. This is what happened with Sudan," the perceptive and preternaturally kind caretaker explained.

Over the long hours we spent with him, he also touched upon several other subjects not directly related to the rhinos' well-being. Zachary proved equally good at reading humans as he was at reading animals. As the jokester of our group, Matjaž was able to relax him with his Pidgin English skits, until Zachary stopped bothering to conceal his tobacco habit.

After that it was official: we were no longer mere professional acquaintances, we were good friends:

"How old are you, Sir?"

"Dirty."

"And how old is your brother, Sir?"

"Dirty too."

"What's the time, Sir?"

"Too dirty."

It may all have been tremendously daft, but the four of us were too busy enjoying ourselves to notice. Our sides hurt from laughing. However, even when convulsing with laughter, Zachary kept one eye on Najin and Fatu. The routine soon evolved into our signature theme, one which all of the Ol Pejeta rhino caretakers knew by heart by the time of our next visit.

"How do you locate a rhino at night?" Zachary demanded with a riddle-me-this grin.

"We give up. How?"

"You follow the farting!" the slight man exclaimed and cracked up.

Even during the day, the rhinos emit gas at an almost frightening rate. At night, the fermented grasses in their stomachs produce a steady barrage of farts so clamorous they can easily wake you up.

Zachary once shared the story of how he fell from his bicycle right in front of a lion. He screamed until he could no longer produce a sound. His life was saved by the bicycle bell setting off while the bike crashed to the ground. The lion just gazed at him funnily for a few moments, then sauntered away.

"I wasn't able to sleep that night," Zachary related with a wide-eyed grin. "And I imagined the lion maybe couldn't sleep either!"

Dusk slowly settled over the rhino graveyard. A cold wind was gathering momentum all over the savannah, and the sky was threatening to cook up a storm. Zachary gazed somewhere far off into the distance. He seemed to be holding his breath, and we could sense what he'd say next would be anything but a joke.

"This whole thing," he eventually exhaled. "It's so very sad. You see, this is what extinction looks like. If the rhinos only died from natural causes, there would still be thousands and thousands left. Human beings are the most savage predator this planet had ever seen. We have done so much evil. The least we can do is to learn from our mistakes."

Unlike elephants, rhinos generally have poor memories. That is probably just as well. If their memory was as good as the elephants', they would take revenge for everything that has been done to them, Zachary added.

Yet despite everything, he remains an optimist. He is convinced that sooner or later, the scientists will succeed at creating a northern white rhino baby. "Najin and Fatu are the last ones, but all of us here have our fingers crossed this will soon no longer be the case. When the next baby northern white rhino is born, there will be great celebration all over the world!"

Keeper's shadow on Najin's back.

Keepers are emotionally very close to the rhinos.

Zachary Mutain and Najin.

Early morning in Ol Pejeta—just out of the stables.

CHAPTER THREE

Poaching

A Long Genocidal Chain

RAZOR WIRE. TALL, ELECTRIFIED FENCES. MILITARY-GRADE OBSTACLE courses. Watchtowers. Rangers being drilled like members of elite special forces. The militarization of conservation. Aggressive intel-gathering activities. Increasing cooperation of local, regional, and state authorities. Constant field presence and strong financial backing from international conservation organizations. A global media campaign for raising awareness. Vast support garnered on social media. Ever-stricter amendments to legislation. Increased surveillance along key transit checkpoints and borders.

None of the above have done the trick. Poachers and organized crime networks always find a new way of slaughtering endangered species and smuggling their remains to wherever there's a market. The whole enterprise remains much too profitable for the villains to just give up. Therefore, bribes and loopholes remain the order of the day.

Poaching in Africa is the perfect embodiment of our global predatory economic system. Each year, the illegal market for wild animals and their products brings in some twenty billion dollars in total. Within the international-crime pyramid, only drugs, weapons, and human trafficking are more profitable.[1]

And so, the sixth mass extinction keeps gathering pace, born of our greed and insanely rising numbers. The rate of extinction for plants and animals is now a thousand times above its historical average—the worst it has been since the demise of the dinosaurs. That means the worst in sixty-five million years.

Vanda Felbab-Brown, the author of *The Extinction Market: Wildlife Trafficking and How to Counter It*, believes this simple fact should have the same impact on our consciousness as climate change.[2]

The sixth mass extinction is yet another global catastrophe dire enough to force us to address the human reasons behind it. After all, the poaching, the illegal wild animal trade, and the horrendous decline in biodiversity also present an ever-mounting threat to public health.

On the Asia-dominated global market, the most sought-after commodities remain rhino horns, African elephant tusks, and pangolin scales. Six of the eight pangolin species are currently endangered.[3]

POACHING: A LONG GENOCIDAL CHAIN

Chinese traditional medicine is approximately 2,200 years old. Most of the ingredients for its remedies are plant-based. The Chinese, however, also use around 1,500 animal species to prepare their pharmaceutical products.

Many of them are endangered, even critically. Twenty-two percent of the 112 most used species can be found on the endangered list. A further 51 percent could soon become endangered, according to Zhibin Meng of the Institute of Zoology at the Chinese Academy of Sciences.[4]

Officially, the sale of endangered-species ingredients (tiger bones, ivory, pangolin scales, rhino horns, etc. . . .) is strictly regulated. In reality, everything is available for the right price. Across Asia, products obtained from wild animals also serve as much-coveted status symbols.

In Chinese medical texts, which date as far back as the fourth century, rhino horns were listed as a remedy for snake bites, hallucinations, typhoid fever, food poisoning and many other ailments. Up until the end of the eighteenth century, rhino horns also saw medicinal use in Europe, for instance at the French royal court.

The mass slaughter of rhinos is therefore nothing new.

Between 1849–1895, during the height of African colonization, 170,000 rhinos were killed for their horns. Most of the "product" eventually wound up in the United Kingdom, where it was used to make kitschy faucets, door handles, and ornaments. Back then, only a small portion of the horns were exported to China.[5]

After World War II, Yemen became the planet's largest market for rhino horns, which were used to produce the famous *jambiya* daggers. In 1980, roughly 90 percent of Yemeni adult men carried a dagger of this type. The home market's demands fueled the murder of at least 65,000 black rhinos, their horns being considered the preferred material for the handle.

1975 saw the implementation of CITES—the Convention on International Trade in Endangered Species of Wild Fauna and Flora. Ten years later, only 3,800 black rhinos were left on the entire planet. The blood trade went on as before. When prices climbed through the roof, the Yemenis took to selling off their daggers to Asian merchants. Once ground down into precious powder, the daggers ended up in Vietnam, Hong Kong, Laos, Taiwan—where rhino-horn sales were successfully prohibited in 1992—and, of course, China. Despite China's 1993 ban on the trade.[6]

During the 1990s, the Asian rhino-horn fever seemed to have simmered down considerably. When that happened, poachers significantly reduced their activities, and the African rhino began to slightly recover. The upturn, however, only lasted a few years. All the wars that broke out in the 1990s and the chaos in the South African Republic delivered the next great blow to the rhino population.

This blow essentially wiped out the northern white rhino.

In 1977, all five rhinoceros species (white, black, Indian, Javan, Sumatran) were included in the annex of the first CITES convention. This meant a total ban on international trade in rhino body parts or related products.

The southern white rhino population began to increase, while the number of northern white rhinos continued its steep decline. As already stated, in 1984 only fifteen were left. The black rhino population also kept shrinking. Over the following two decades, the black rhino went extinct in eighteen African countries.

2011 saw extinction of the western black rhino, a subspecies of the black rhino. Poaching was the cause. Both the Javan and the Sumatran rhino are critically endangered. At the time of writing this book, only seventy-five Javan rhinos were left, and less than eighty of their Sumatran counterparts.[7]

In Vietnam—the largest rhino horn market in the last two decades—rhino horn powder has long been instilled with a cult-like, almost mythological status. It is believed to cure everything, from severe hangover to various forms of cancer, AIDS, epilepsy, measles to . . . Covid. It is also used in jewelry production and as an aphrodisiac. It serves as a very popular addition to alcoholic beverages.

Vietnamese law does not prohibit the possession of rhino horns. Only their trade.

In 2003, criminal elements sniffed out an important loophole in the South African Republic's legislation on trophy hunting. The loophole allowed foreigners to export trophies, like horns or entire rhino heads.[8]

The law stated that the first to fire at the rhinoceros in question had to be the person whose name was listed on the permit. The same person was the only one licensed to export the trophy. Suddenly, a great number of Vietnamese started applying for the permits. They were more than happy to pay up to $85,000 for the opportunity to murder a rhino, usually the southern white kind.

2009–2015 was a time of exceptional dread for the rhinos. Over this interval, more than half of all hunting permits in the South African Republic were issued to Vietnamese citizens. The costs, naturally, were covered by their superiors—namely the international crime syndicates, who worked in close contact with corrupt elements in the South African Republic's system.

The most notorious of the criminal gangs was led by the infamous *bon vivant* Chumlong Lemtongthai from Thailand. His criminal empire included several dozen prostitutes. Some of them played a very important role in the rhino trade.

Since the call girls were provided with South African Republic residence permits, one of them only had to appear at the killing site, pick up a rifle in front of a camera, and then hand the weapon over to a hired professional hunt master, or to one of the employees at the farms, where local farmers mostly bred southern white rhinos.

Lemtongthai was arrested in 2011. In 2012, the South African Republic sentenced him to forty years in prison. Following the appeals filed by his lawyers, the sentence was first commuted to thirty years, then to thirteen years. In September 2018, he was released after having served only six years.[9]

Lemtonghtai was a subordinate of Vixay Keosavang from Laos, the past millennium's foremost trader in rhino horns and other wildlife products. In her book *Poached: Inside the Dark World of Wildlife Trafficking*, Rachel Love Nuwer dubbed him the Pablo Escobar of the wildlife trade. After extensive research, Nuwer was able to establish that most of the relevant countries simply do not bother with breaking up the smuggling chains and protecting the rhinos.

The background of the South African Republic's blood trade was meticulously eviscerated by the investigative journalist Julian Rademeyer in his brilliant book *Killing for Profit: Exposing the Illegal Rhino Horn Trade*.

Among other things, Rademeyer proved that the rhino horn trade involved the complicity of several employees of the Vietnamese, North Korean, and Chinese embassies. Following Rademayer's findings, the South African Republic's authorities slightly updated the legislation. Under the new decree, only "sport hunters" were able to obtain the permits for killing rhinos.

The Vietnamese were thus blacklisted, but they were swiftly replaced by the Czechs. The home country of the Dvůr Králové Zoo also had a strong Vietnamese diaspora, which dated back to the 1970s. The smuggling network was quick to establish contact.

The Czech hunters were subcontractors of Asian cartels, which used the same routes for smuggling wildlife products as for smuggling guns and drugs. "It's called 'pseudo-hunting,'" Julian Rademeyer wrote. "It isn't hunting, it's shooting. And the 'hunters' are little more than pawns recruited by criminal syndicates to acquire horn for the medicinal black markets of Southeast Asia and China. Everyone knows it, from the out-fitters and professional hunters who arrange the hunts to the permitting officials and bureaucrats who are meant to enforce the regulations."[10]

The illegal wildlife trade in the South African Republic involved many farmers, veterinarians, professional hunters, accountants, business-men, public officials, members of various security outfits, and politicians. As Rademeyer points out in his deeply researched book, "Vietnam, today, is without doubt the world's leading destination and consumer of rhino horn. The rot that has allowed the illicit trade to flourish extends deep into the Vietnamese government and officialdom. There is no political will there to end the trade. Arrests and seizures at Vietnam's ports occur rarely, if at all. And Vietnamese diplomats still appear to be involved in the trade despite the scandals in South Africa."[11]

According to John Lukas, the president of the International Rhino Foundation, his organization has opted for a twin-pronged approach to fighting the rhino horn trade. The two main priorities are to contain poaching and to disrupt smuggling chains that lead from Africa to Asia.

"In cooperation with our local partners, we're very active in Vietnam and in China," Lukas told us over the phone. "The African wildlife is mostly killed with firearms. Rhinos are always the main targets. They are killed on sight even when the poachers are hunting elephants. Rhino horns are simply too precious, you see. In Asia, hunting rhinos is based on the laying of traps. The animal gets caught and dies. When the poachers find the carcass, they cut off the horn. And the rest is left to rot."

John Lukas was talking to us from Florida, where he had been employed by the Jacksonville Zoo to develop conservation programs and forge strategic links with other conservation initiatives.

"What I really want to stress," the conservation legend continued, "is that over the last thirty years, the poachers hadn't killed a single Javan rhino! Instead, they were killed by the vanishing of their primary living environment. The cutting down of forests and other incursions into the rhinos' life space have made them disperse over a very large area. And so they began to lose contact with each other. The mating prospects get slimmer every year. Which is also the case with the Sumatran rhinos, though to an even greater extent! The rhinos are still seeking, but they are no longer able to find each other. This is extremely tragic."

In Lukas's view, the main problem behind rampant poaching is still politics. The only realistic possibility of shutting down the rhino horn trade in Vietnam and China is from the top down.

International crime syndicates have long set up a highly functional system of comprehensive logistics, including bribery. "Through their networks, they're able to move huge quantities of illicit cargo," Lukas explained. "To switch from guns, drugs, and human trafficking to rhino horns, they only needed to tinker with the already existing system. These criminal organizations are extremely wealthy. They have their own armed units, helicopters, informants, and lawyers. They're extremely dangerous and they are easily able to pay people to do their killing for them."

Over the past few years, John Lukas became thoroughly disgusted with the South African model of "rescuing" the rhinos. Under the banner of saving an endangered species, private farmers took to breeding southern white rhinos, black rhinos, and lions for trophy hunting. The set up was perfectly legal and brought tremendous revenue and power to criminal syndicates.

"It was just awful!" Lukas snarled. "Some farms were regularly cutting off their rhinos' horns and then they would stockpile them in warehouses, waiting for the right market conditions. Which of course only further fueled the rhino horn fever and expanded the black market. What sort of a message is being sent to the world by legalized slaughter of an endangered species? Shouldn't we treat every single one of those animals as special? Shouldn't our task be to protect them, not to kill them?"

At the time of writing, only around 30,000 rhinos—of five different species—are left on the planet. Each year, the poachers claim another thousand. Ninety-eight percent of all African rhinos—white and black—are located in Namibia, Kenya, Zimbabwe, and the South African Republic. The latter country is home to a little over 20,000 rhinos, or two thirds of what is left.

According to IUCN data, two of the three Asian rhino species are critically endangered, namely the Javan and the Sumatran rhino. The Indian rhino, however, is vulnerable by IUCN standards.[12]

With the commendable exception of Kenya, the global war on rhinos is just about reaching its final phase.

"Rhino horn, reptiles, guns, drugs—they are all commodities to be bought and sold on the black market, and the smuggling routes are often the same," investigative journalist Julian Rademeyer described the situation.

"Killing poachers doesn't achieve anything. There are so many poor guys out there and criminal elements that are prepared to take the risk to make quick bucks. No matter how many of them you shoot or arrest, you'll never stop it. The only way is to cull the market. You have got to get to the guys at the top," Julian Rademeyer was told by Blondie Leathem, a veteran of Zimbabwean bush war and, later, a ranger.[13]

There needs to be at least a credible threat that poaching will be punished. So far, this threat has been perceived as minimal; for the most part, rightly so. The two main reasons behind this sad state of affairs are corruption—especially the direct links between the poachers and the authorities—and an absolute lack of general concern.

The global poaching crisis exists alongside a vast and still expanding legal trade, American researcher Vanda Felbab-Brown writes. In her opinion, the legal trade sometimes ensures the protection of the environment that the animals are culled from, but for the most part only contributes to the laundering of illegal activities.[14]

The key to successfully tackling the poaching problem both locally and globally is the inclusion of the local communities in conservation projects. Most of the poachers, after all, hail from said local communities. Financial aid is simply not enough. According to Vanda Felbab-Brown, the strategy of cooperating with local communities to benefit critically endangered species should also involve the psychological, legal, and political empowerment of said communities.

Beside rhinos and pangolins, poachers are also endangering another iconic species: elephants.

Ivory use is as old as human civilization. Around 35,000 BC, the peoples of what are now Russia and Germany were already shaping ivory into tools and jewelry. Approximately in 2,750 BC, the Egyptians launched a rapacious march on the Sahara and exterminated most of the elephants. On the other side of North Africa and in Asia Minor, the Romans were prepared to kick off their own killing sprees. By the seventh century, no elephants were left in these areas. Ivory also saw wide use with the Assyrians, the Babylonians, and the Phoenicians.

The next wave of elephant mass murder came during the colonization of Africa. The European enslavers were not the only ones responsible. To an even greater extent, the blame was on their Arabian counterparts.[15]

Between 1500 and 1900, Africa's elephant population fell from twenty-six million to ten million. In the first half of the eighteenth century, the ivory trade found its hub in Zanzibar. Over the century's second half, approximately 24,000 pairs of elephant tusks were shipped from there each year. After World War I, African elephants were hunted by the British army. Its airplanes were used to kill the elephants, and then the infantry would sweep in to remove the tusks.

The Americans were also far from immune to the world's obsession with ivory, though for the most part as buyers. Over the first two decades of the twentieth century, they purchased some 600,000 pairs of tusks. Ivory was put to many new uses, for instance as the material of choice in making piano keys. Nothing short of a genocide was perpetrated on the elephant.[16]

After World War II, when the shooting expeditions of the rich whites gave way to nature safaris, the number of Africa's elephants slowly began to stabilize.

According to Iain Douglas-Hamilton, the founder of the Save the Elephants organization, some 1.3 million forest and savannah elephants lived on the continent in 1960. In those days, Douglas-Hamilton recalls, no one's wildest nightmares included the notion that one day, intruders armed with automatic rifles would break into the parks to murder elephants and rhinos.

Yet that was precisely what happened.

The 1970s and the 1980s were a time of great massacre. Over this period, approximately 7,000 tons of ivory was exported out of Africa.[17] The toll was about 70,000 dead elephants per year. The price of ivory skyrocketed. In 1970, one kilogram was worth around $5; two decades later, the price was almost $200. Poaching became a lucrative global enterprise with Hong Kong as its seat.[18]

The largest end buyer of ivory was Japan, the economic miracle of that era. The Japanese bought up more than 40 percent of all available ivory and used it for the manufacture of customized personal stamps. Back then, the international ivory trade was still legal. Conservationists now refer to the period as "the first elephant holocaust."

The fact that the set up was legal does not mean there was no poaching. Quite the contrary. Iain Douglas-Hamilton reported that in 1973, only 44 percent of Kenyan exports were of legal origin. The rest was obtained through poaching in Somalia, Uganda, the South African Republic, Sudan, Angola, and the Central African Republic. Much the same went for the supply of rhino horns.

At the end of the 1980s, even greater chaos engulfed Tanzania—a chaos stoked by organized crime and rampant corruption. An investigation revealed that illegally obtained ivory formed as much as 90 percent of the country's exports. In Kenya, where authorities and criminal syndicates had forged the tightest links, 63,000 elephants were killed during 1979–1989. That's three quarters of the entire population! Some of the rangers and the security staff were also involved in the bloodbath.[19]

In 1988, 88 percent of all ivory imported by China was obtained illegally, through poaching. If the bodies were human, we would call it a genocide, Iain Douglas-Hamilton wrote at the time.

To put it mildly, the legal quotas prescribed by CITES failed to produce the desired effect.

Between 1979 and 1989, the number of African elephants fell from 1.3 million to 600,000. Only then came the introduction of bans on ivory imports from specific countries. Following market demand, the wholesale murder of elephants was in decline. But the economic rise of China, with its hundreds of thousands of *nouveaux riches* and their obsession with status symbols, ushered in the next wave of slaughter.

By 2002, China was the world's biggest ivory importer, and the chief illegal ivory trade facilitator.

Once again, poaching was on the rise. After 2010, the entire enterprise exploded—due to the wildly increased demand, the malfunctioning legislation, the corruption, the rise of new criminal organizations and especially the enormously strengthened political role that China played in the most exposed African countries.

In Mozambique, the elephant population was halved in just five years, between 2010 and 2015. Much the same went for Zambia, which became virtually a Chinese colony.

In Tanzania, the land of safari and Kilimanjaro, the number of elephants fell by 60 percent in five years.

In Chad and West Sudan, the remaining elephants were killed by the Janjaweed—members of the infamous Arabic mounted militia, armed by the Khartoum authorities to combat the indigenous African peoples in Darfur.

This was the twenty-first century's first genocide. It was first perpetrated on human beings, then on elephants. The hunting season was open. The body count grew unimaginable. The elephants vanished. The people who managed to survive the carnage have spent the last decade and a half in refugee camps.

The Great Elephant Census, taken between 2013 and 2016, determined that approximately 144,000 African elephants were killed between 2007 and 2014. This means that their population fell by a third over seven years.[20]

The quotas were thus rendered completely irrelevant. Even worse, according to numerous conservationists, the quotas had accelerated elephant poaching. Timothy Tear, executive director of the Wildlife Conservation Society's Africa Program, says it is virtually impossible to discern between legal and illegal ivory.

"The problem is you can't distinguish between legal and illegal tusks. Heroin and human trafficking—that's easy to distinguish because it's all illegal. The situation with ivory, however, is more similar to arms trafficking, because some weapons are legal, others are illegal. But even arms have serial numbers," Tear was quoted in Rachel Love Nuwer's book.

In 2017, China—which was already importing ivory from the Arab world as far back as the second century—finally banned the trade. But this just meant the business was immediately relocated to Vietnam and Laos.[21]

For Keith Somerville, author of an influential book called *Ivory: Power and Poaching in Africa*, the resurgence of the ivory trade can be chalked up to a mixture of corruption, wars, and poverty in Africa, as well as the price hikes driven by the heavily increased demand for ivory on other continents: especially in Asia.

Despite occasional confiscations and arrests of smugglers, Somerville believes Chinese authorities have done next to nothing to prevent illegal imports from reaching their domestic market.

In the context of China's economic success and its rapidly increasing African presence, all the above have facilitated the formation of a brand-new market with a brand-new supply chain. On return from temporary assignments in Africa, Chinese workers are able to smuggle in enormous quantities of ivory and earn a great deal of money.

This is part of the reason why China slowly took over a great share of the market from much more established players. Then came the impact of the global pandemic. It encouraged poaching in numerous regions all over Africa.

At this stage, what can even be done? How can the Chinese market or any other large ivory market even begin to be shut down?

Frank Pope, the CEO of Save the Elephants, has been tackling the problem for over twenty years. "China's size, economic clout, and political influence form a fatal combination," the former BBC star told us in Nairobi. "Especially in combination with its long tradition in ivory trade. This is a recipe for trouble. We've been focusing on raising awareness in China for a while now."

Pope is a veteran of the international conservation cause, having testified on the perils of the ivory trade before a U.S. senate committee as far back as 2000. As an example of sound practice, he likes to bring up Kenya's Samburu National Reserve, which we visited in the spring of 2021. A detailed report of the visit will be provided in chapter 8.

"We sought out Chinese opinion leaders and celebrities willing to spread the word about the good work being done at Samburu. We brought some of them over to Kenya—for example the basketball player Yao Ming, who is incredibly influential in China. Yao contributed a great deal to the Chinese ban on the ivory trade being effected. We also invited a few of the most influential Chinese journalists. For a long time, covering the ivory trade was forbidden in China. This is no longer the case," Pope broke down the recipe for educating the public.

"Yet the international conventions, including CITES, clearly do not work!" we challenged his optimism.

Pope winced. "In order to really understand the CITES convention, you have to understand how it came to be," he replied diplomatically.

"It originated as a mechanism of control over the endangered-species trade," he elaborated. "The convention was not signed by conservationists. It was meant as a trading mechanism: a form of market regulation agreed upon by the countries which signed it. This needs to be well-understood before we accuse the convention of being ineffective. There is certainly no lack of these sorts of accusations and appeals to reform the convention."

What, then, is the point of the convention that many in conservationist circles find highly problematic?

"Well, to you and me, some of the things enabled by CITES may seem revolting. For example, the export of wild animals to zoos. To my mind, very few things are more disgusting than that. But you must realize—in no way did CITES negatively impact the elephant population of Africa. While weighing the convention's pros and cons, you must be brutally rational. The decisions tied to the convention are being made by science. This basic set-up needs to be preserved—protected from many detractors who are losing faith in the convention. CITES is there to enable control over trade. Just imagine a world without these limitations! You see, sometimes we have to be very careful and back the lesser evil," the Save the Elephants leader warned.

Frank Pope eventually switched the topic from elephants to rhinos. After all, the fate of the two magnificent species has long become intertwined.

"The rhinos are still subjected to everything we fought against while trying to protect the elephants," Pope related. "The horns are simply too pricey for the rhinos to survive. And I mean this very literally. In my opinion, the rhinos are no longer capable of surviving anywhere in Africa without round-the-clock armed protection. There are simply not enough rhinos left to meet the demand in the Far East. If the ivory trade were still legal, exactly the same would be the case with the elephants. They could only survive in fortresses. In the South African Republic, where hunting is legal, they've even been breeding elephants and rhinos for hunting purposes and *still* there is a great deal of poaching going over there! These animals, however, can no longer be described as 'wild.' They are bred. It is a horrendous thing."

Raoul du Toit, who lives and works in Zimbabwe, is regarded as the man who contributed most to the survival of the black rhino. Over the last few decades, his initiatives and field work effected tremendous change in Zimbabwe, Zambia, Namibia, and many other places.

Du Toit is also renowned as one of the advisors of the International Rhino Foundation. Regarding the project of saving the northern white rhino, he was involved as a technical member of the African Rhino

Specialist Group, functioning within the framework of the IUCN's Species Survival Commission.

Appropriately enough, the sturdy and bearded legend of African conservation has received countless awards in his chosen field. In December 2021, we called him in Harare and inquired about the news from the front lines of the war against poaching.

"The most important news lately is that there's not a great deal of bad news in the media," du Toit responded. "But, though the media have hardly reported on it, the worst news are currently coming from Botswana. Over the last few years, many rhinos were moved to the said country. The politicians bragged about it, the term 'success story' was being bandied left and right. But that initiative relied on a very militaristic approach and unfortunately professional poachers tend to do a more efficient job of killing rhinos than professional soldiers do in killing. . . ."

Talking to us over Skype, du Toit paused to phrase his next words carefully. "Following the regime change, the conservation approach became even less holistic. Now there's a lot more poaching. A disaster is taking place, though we still don't know its real extent."

We could sense the pain Raoul du Toit had to deal with when discussing such calamities for his beloved animals.

"When the poaching increases so dramatically, it usually sends a positive message to the criminal syndicates," he went on. "The message is that the other rhino populations in the region could now also be easy pickings. The South African Republic still sees a lot of poaching, but at this stage less than it used to. You may have heard that, over there, thirty rhinos were recently killed in a single day. At the Kruger National Park, they chopped off the horn from a southern white rhino who was still alive! He died abandoned and in terrible agony. In Namibia, things have turned for the better over the last few years. Zambia is questionable. Many of the poachers reach Botswana through Zambia. We can only hope that what's happening in Botswana won't impact the rhino populations in other Southern African countries."

As we concluded our Skype session, du Toit expressed great concern at the fact that so many things were happening at once. He feared all the chaos might herald the next great wave of rhino butchery. He also reiterated his long-held view that the poachers were mostly cannon fodder for the big players, who almost always get away with it in the end.

For a good long while, Raoul du Toit hasn't been directly involved with the physical safety of rhinos in their natural environment. The focus of his activities has long shifted to managing rhino populations.

The key, he says, is understanding what is happening to the rhinos, how they react to the changing environment that humans induce for them through poaching or conservation counteractions such as trans-locations, and how the poachers are organized. Gathering reliable data is absolutely essential. This is something most of the people we talked to over the last two years agreed upon.

"You see, it's not enough to send armed men against the poachers. When the culprits are located, it is usually already too late. Therefore, proactive prevention is the most important thing. This is where me and my colleagues have placed a great priority. We've been focusing on secur-ing the financing for the intelligence system and on studying existing databases. Our regional network is already pretty strong. A lot of data is being exchanged. But the circumstances in each country are highly specific. The same goes for the level of coordination between various countries."

When it comes to the fight against poaching, Namibia serves as an example of sound collaboration between all the key players—public, non-government and private. The coordination in Zimbabwe and South Africa is less effective, du Toit told us. In Botswana the system has com-pletely fallen apart.

Du Toit is convinced that when discussing the poaching scourge, both hopeful news and examples of existing sound practice should be emphasized. In his view, the story as it has been unfolding recently is not a total catastrophe.

"In Kenya, the poachers didn't kill a single rhino in almost two years. Something quite similar could be said for Zambia. In Zimbabwe, we only lost five rhinos over the last twelve months. The overall number of rhinos is slowly rising. I believe it will continue to do so in the near future. Some successful relocations have also taken place. For example, the relocation of black and southern white rhinos to Rwanda. We also managed to build up a new population in Zimbabwe. It's not all bad," the Zimbabwean conservationist explained in a reassuring tone.

In de Tout's opinion, we would do well not to separate the international illegal wildlife trade from the other main branches of organized crime.

The poachers and the smugglers are usually just details in a far larger picture. For the most part, they are mere subcontractors for large smuggling conglomerates, which also deal in guns, drugs, people, and plenty more "products" besides.

"In Zimbabwe, the demand for rhino horns experienced a significant fall over the last few years," de Toit related. "There are of course several reasons for that. Some criminal activities are much less risky than selling rhino horns to the Asian market and equally profitable. The illegal gold trade would be a good example. In Zimbabwe, this is a huge industry. It is wrong to say the value of gold equals that of horn powder. You know, that's nothing but a great myth," du Toit continued.

Among other things, a recent study of rhino horn use in Vietnam demonstrated that the price of rhino horn dust is no longer as inelastic as it used to be.

The study, involving 300 Vietnamese end users, determined they were prepared to pay only up to a certain amount. This means that a sizeable price hike could significantly reduce the demand. On the other hand, if the price drops, the demand—and the incentives for poaching—would certainly increase. In Vietnam, rhino horns are still regarded as premium luxury goods.

The study's other important finding stated the following: Vietnamese consumers are only interested in the horns of the rhinos living in the wild, and not in the horns of the rhinos being bred on farms in, say, the South African Republic.

According to Raoul du Toit, this clearly demonstrates that the potential legalization of the rhino horn trade so vigorously pushed for by the South African breeders and certain other lobbyists would fail to put a stop to poaching.

The study also laid waste to the argument that legalized trade could be a means of combating illegal trade and poaching. That is simply not true, du Toit is convinced. The study, which also highlighted a total lack of Vietnamese consumers' moral reflection on the origins of the rhino horn powder, confirmed many of his firm beliefs.

When pressed for news from China, du Toit replied there was surprisingly little. In his opinion, this could mean that the Chinese authorities' recent efforts to step up the prosecution of some of the smugglers might have somewhat reduced the demand, after all.

"Both the open parks and the closed, guarded reserves need a lot of money. And also, political support and close collaboration with the international organizations, which can offer their expertise and sound management practices. If the governmental staff is not properly motivated, this can open the door wide to poaching," Raoul du Toit was clear.

He went on to assure us that climate change and the resultant acceleration of wildlife extinction had made even the world's private capital realize that huge investments have to be made into the natural environment as they are, themselves, a form of capital.

If, of course, we wish to preserve them.

CHAPTER FOUR

God's Particle in Human Hands

THE TRANQUIL AND EASY-GOING ROUTINE AFFORDED TO NAJIN AND Fatu at the Ol Pejeta Conservancy is disrupted three times each year. Approximately every four months, Dr. Thomas Hildebrandt flies in from Berlin, in full veterinary regalia and brandishing his magic Sport Billy bag. It's crammed with the tools needed to perform gynecological procedures on the planet's second largest mammals.

The world-renowned veterinarian and honorary scientific father to countless endangered animals is the main protagonist in the tale of saving the northern white rhino.

On mornings when Najin and Fatu see him approach their pen, they always get restless. Even the staunchest rhino aficionado would be hard-pressed to endow the species with an especially brilliant memory. However, upon catching a glimpse of Dr. Hildebrandt, "the girls" are instantly aware that this will not be a day for munching grass, silent communion with their peers and caretakers and drowsing in the deep shade.

What is it that rhinos dream of? We couldn't say. But for Najin and Fatu, Hildebrandt's arrival is like a nightmare that keeps coming back. It is an ordeal they must undertake for the sake of their practically extinct species.

As a specialist in artificial fertilization, Dr. Hildebrandt has been collecting the pair's oocytes (immature eggs) since 2019. Once they are collected, they are swiftly delivered to the embryologist, Cesare Galli, in Cremona, where he creates northern white rhino embryos.

The frequent narcosis makes the program of oocyte collection very strenuous for Najin and Fatu, both physically and psychologically. Therefore, in November 2021, the BioRescue team decided to retire Najin from the ordeal, as important as the project's success might be.

Sedating a two-ton rhinoceros is far from an easy task. For Najin and Fatu, the process is even worse. Everything connected to the collection of their oocytes is highly unpleasant for them, even torturous, including the period after they are roused from artificial slumber.

At the beginning of 2022, Dr. Hildebrandt was charged with another task—one he had been waiting a long time to fulfill. We can only hope that it isn't too late. As the go-to expert for transferring southern white rhino embryos into surrogate mothers, he will be the one to transfer the first northern white rhino embryo into a surrogate southern white rhino mother. By the time you read these lines, he will perhaps already have done so.

When—or rather if—that happens, Hildebrandt and his dedicated team will become the saviors of an almost extinct species.

As a pioneer in the field of the assisted reproduction of large mammals, Dr. Thomas Hildebrandt has already "fathered" the next generation of several endangered species. As the head of the department for Reproduction Management at the Leibniz Institute for Zoo and Wildlife Research and the chair of Wildlife Reproductive Medicine at the Freie Universität Berlin, his expertise, experience, and dedication proved invaluable to the project of saving the northern white rhino.

No one is more impatiently awaiting the birth of the first northern white rhino from a lab-created embryo, delivered by a surrogate southern white mother.

Thomas Hildebrandt dreamed of being a veterinarian since he was six years old. Born in 1963, he spent his childhood wandering the forests and meadows of East Germany. He wanted to help every sick or wounded animal he came across.

He was always fascinated by the concept of biodiversity. He wanted to understand how the animals differed from each other, how evolution had put them together and shaped their behavior. In short, he wanted to know what made them tick.

"This childhood obsession wasn't a means of escaping human society," he clarified. "It was simply that I was always mesmerized and awestruck in the presence of nature."

As he matured, his feelings didn't change one bit.

"I was born to a family of engineers. Which made me something of an outcast. A veterinary career certainly wouldn't have been their choice for me. However, they never tried to stand in my way," Hildebrandt reminisced on his formative years during one of our several interviews.

Even in high school, he spent his summer vacations working at the East Berlin Zoo. "If you wanted to become a vet in the German Democratic Republic, you really had to excel in school. I tried to the best of my abilities; I was getting excellent grades, but it wasn't enough, since I wasn't deemed politically appropriate," he recalled.

In the early 1980s, a time when the East German communist regime was at its most oppressive, Hildebrandt often wrote articles for various school papers, openly pushing for a freer and more democratic society. He also firmly refused to become a member of the communist party. So, he was notified in advance that he would not be allowed to become a veterinarian.

"When I was still very young, I often came into conflict with the Party. They were furious that I wouldn't join. My application for studying veterinary medicine was turned down. I was branded as an enemy of the working class. A member of the bourgeoisie, no less! So, after high school, I was forced to start working at a dairy farm milking cows," he related with a distinct twinge of bitterness in his voice.

After his stint at the dairy farm, he got a job at a veterinary school, but only as an unqualified worker. "I actually had a lot of problems securing even such a lowly position—given the contents of my file, which I was never allowed to see."

In a way, his communist persecutors were right. Even as a young man, Hildebrandt burned with the desire to change the world. "Only if you doubt, pose questions, actively seek solutions, and step out of the mainstream can you effect positive change," he remains convinced to this day.

We wondered, was he trying to transfer the *embryo of freedom* into the un-free communist world?

"Well, I guess you could also put it like that," he laughed in response. "I really wanted to help empower the common people. Because you see, all the power in the German Democratic Republic had been seized by the regime. The people charged with running the system weren't the best and the brightest. No, they were the most compliant holders of party membership cards!"

With "Compliance" not being his middle name, young Thomas took his lowly position at the veterinary school very seriously. He cleaned up the entire school building, which was over a hundred years old and kept inventing new hygienic procedures as he went along.

His zeal impressed one of the professors, who backed Thomas's application to the university. After a five-year wait, he was accepted in 1986, roughly when *perestroika* and *glasnost* were taking root in the Soviet Union.

Though he was still not a party member and had also refused to do his mandatory military service, he was chosen as the best student in his first year of studies. He also received several national awards. As an outstanding student who already began preparing his doctorate during his first year at the university, he was sent to the Soviet Union. Before that, Hildebrandt spent a few months on Mauritius, where he worked with monkeys, flushing out embryos which didn't take during the transplant. He also performed ultrasound diagnostics. The experience was a major turning point in his life.

But a turning point was also on the cards for his home country, when in the fall of 1989, the Berlin Wall, communism, and the entire Eastern Bloc fell.

"My experience with the Berlin Wall is rather curious, given that the wall had cut our family in half. The part of the family which remained in the Federal Republic of Germany was allowed to visit. Whereas those of us who'd fallen under the German Democratic Republic were not allowed to go anywhere. So, I guess we were visited like animals in a zoo are visited," Hildebrandt laughed at his own interesting analogy.

"At all times, we were aware of what was going on beyond the Wall. We were ready to embrace change. When the Wall was brought down, hundreds of thousands of East Germans immediately headed westward. I, on the other hand, began driving to work even further to the east."

Back then, Hildebrandt was involved in an important project in the north of East Germany. "I was implanting female yak embryos into cows. The procedure was quite similar to what we're now trying to do with the northern white rhino embryos. The difference is that back then we had to do a lot of experimenting. You know, Mongolian cheese made from yak milk is simply exquisite. We were hoping to start producing it with the help of said procedures. Two of the cows got pregnant, but the project was ultimately a failure, since the yak placenta turned out to be incompatible with the cow uterus," Hildebrandt recalled.

This was the tale of how he spent six weeks driving in the opposite direction of the masses fleeing to the west, where every new arrival was routinely handed out 100 deutsche marks. "The eagerness of some of them still strikes me as morally objectionable and rather bizarre. You see, the people who first broke through the holes in the Wall were riddled with members of the communist party. I knew a number of them personally. They were the ones who'd denied us our freedom for so long and then they seized the very first opportunity to enjoy life!" the German scientist mused on the great historic turning point.

The fall of the Berlin Wall spelled great change not only for Hildebrandt's home country, but for the entire world. The end of the Cold War and the birth of a new world order caused tectonic shifts all over the globe. Not all of them were positive. Countless new conflicts were set in motion, and countless already existing ones were exacerbated.

Many of them took place on the African continent, the home of the northern white rhino.

Hildebrandt's path became destined to cross with rhinos at the beginning of his veterinary studies when he chose embryo transfer as the subject of his doctorate. At the local zoo, he successfully transferred an embryo of an endangered okapi into a goat.

Every year, "head-hunters" from the East German Academy of Science would visit his university. They seemed very enthusiastic about the young scholar's forward-looking ideas. In 1985, Hildebrandt performed the first ultrasound scan on a rhino at the East Berlin Zoo. The usefulness of this non-invasive diagnostic method, with no harmful consequences to the examinee, delighted him no end. This was his first contact with the endangered horned giants, who went on to profoundly influence his life.

Since the reproductive organs of the female rhino could not be accessed "manually," Hildebrandt decided to develop an ultrasound-based tool for the examination of rhino uteruses. With a few adaptations, the device could also be used for oocyte collection.

Hildebrandt developed the same plan for elephants, his favorite animals. Quite a *handy* solution, we joked. "It *was* rather *handy*, yes." Hildebrandt joked back. "After each elephant examination, I immediately had to take a shower."

Working with okapis made Hildebrandt realize the velocity of the man-made sixth mass extinction wave.

"The okapis are highly vulnerable to tuberculosis," he explained. "Sadly, the animals who got sick had to be put to sleep. Back in 1987, we didn't yet have the option of storing their genetic material, so they were gone without a trace. That really made me think. I wanted to create a system for preserving their genes. For example, through oocyte collection, embryo transfer and, of course, also stem-cell technology, which we are exploring in collaboration with our Japanese colleagues. We discovered these technologies could help us revitalize the genetic material of deceased animals."

Hildebrandt patented a few procedures and devices, which have already been put to good use in the assisted reproduction of elephants, giant pandas, Asian lions, and tigers.

When coming up with these procedures and devices, Hildebrandt closely collaborates with his brother, who is an engineer. The two of them recently devised a system of cages preventing the *proteus* from devouring their own young. *Proteus anguinus*, or "the human fish," is a species of cave-dwelling aquatic salamanders. Hildebrandt's innovation was quite a coup for the reproductive prospects of the proteus living at the Hermannshöhle underground laboratory in Germany.

Thomas Hildebrandt agreed with our assessment that the said innovations reflected his family's engineering background. However, he made sure to shrug off any claims of supremacy.

"I have always been blessed with the company of pioneering minds, certainly!" he smiled gratefully. "True, our solutions at the Leibnitz institute are usually the most pragmatic and reliable, but I really wouldn't want to claim authorship for any of our successes. The group is always stronger, more important and more successful than the individual. The community is stronger than any single mind. What I am is someone who helps projects progress and evolve. You need to keep trying, you need to make mistakes, you need to make educated guesses and you need to double check all the time. You need to strive to be the one person who can occasionally inspire the entire group to press on."

Hildebrandt paused briefly. "You know, it's so wonderful to have colleagues who think like you up to a point, but also have their own well-argued views. It's especially great when the teams are assembled from different backgrounds and various areas of scientific research. This is a part of what drives me on. And why I am totally in love with my job."

During our first conversation—it was a magazine interview following our first visit to the Ol Pejeta Conservancy—Hildebrandt readily admitted that he very much enjoyed working with rhinos. However, elephants remained his one true flame. He fell in love with them during the short while he spent working in the United States during the mid-nineties.

Having already created roughly fifty baby elephants with the help of artificial fertilization, Hildebrandt's whole diction changes when the subject turns to the species. "The elephants can read your thoughts, you know, and they can respond in a fashion that will shake you to your very core. It's hard to believe how smart they are. We're talking about a whole other category of intelligence, compared to the rhinos. The rhinos can recognize you; they can accept and tolerate you, but. . . ."

Hildebrandt trailed off, then quickly reined himself in. "All right, if this same question of intelligence was posed to Najin and Fatu's caretakers, you'd probably get an entirely different answer. Those guys spend more time with them than with their families, and whenever I visit, the rhinos are asleep most of the time."

"In your profession, is it possible to treat all animals equally?" we asked the German scientist. "Or are some of them more equal than others? In such a sensitive line of work, how are your emotions best handled?"

From his reply, it was instantly clear that he had already given the subject a lot of thought.

"I mostly work with critically endangered species," he nodded, "which means the first selection had already been done, and these species' vulnerability of course influences my—our—approach. But we need to work hard to keep our focus and stick to our chosen goals. We really can't afford to make a lot of mistakes. So, my answer is simple. No, as a scientist I do not and cannot have the same attitude toward all species. We need to leave our emotions at the entrance to the lab."

Dr. Hildebrandt spoke with the calm authority of someone whose team had long learned how to create new life. In this pursuit, he is still guided by his lifelong wish to help shape a better world. Yet he is also increasingly driven by the desire for humankind to correct at least some of its horrific mistakes. Like many other functionally extinct species, the northern white rhino was integral to entire ecosystems.

"All my energies are devoted to endangered-species conservation and to a gradual stabilization of ecosystems," Hildebrandt related. "I am profoundly gladdened and energized by the fact that the younger generations seem to be much more interested in the planet's future than the ones before. The young also have much more respect for biodiversity."

Hildebrandt's fervent hope is that his work and the work of his team shall not be misused as an alibi to continue with old devastating practices, as a sort of indulgence. "It would, however, be really stupid not to use the reproductive technologies and the tools we've already developed for bolstering the species inhabiting the most endangered ecosystems—for example in a large part of sub-Saharan Africa," the veterinarian clarified. "You see, these ecosystems are home to some of the viruses most harmful to humankind—from HIV to Ebola."

In Hildebrandt's view, climate change, the shrinking of natural habitats, and the intensification of conflict between the world of humans and the world of wild animals all exponentially raise the chances of virus outbreaks. "The role of our fight to save the northern white rhino is decidedly undervalued in this context," Hildebrandt likes to warn again and again.

From the very first stages, the way the world chose to handle the recent epidemic has been very frustrating for the German scientist. "We make drastic decisions, we're wasting unconscionable amounts of money, we're radically altering the world around us, but we only seem to be interested in the short term! Our efforts should be directed at preserving biodiversity. Only in this manner will we be able to protect ourselves from far deadlier viruses."

His views on this were very clear. "We are extremely aggressive in encroaching on animal habitats. We're literally losing our distance from the animal world. We're feasting on everything we can sink our teeth into. This is probably how we came into contact with the Coronavirus in Asia. I repeat, the conservation of biodiversity is absolutely paramount! Yet we're not even discussing the subject, let alone addressing it. The politicians, the media, and the public seem perfectly happy to ignore it. The solution is very simple. We must learn to respect nature, or we will be gone."

Dr. Thomas Hildebrandt sees the rescue of the northern white rhino as one of his last major tasks before retirement, and also as the key project for saving the critically endangered species of the future. For a number of years, he has been devoting most of his time to the fate of the northern white rhino. Time, after all, is precisely what Najin and Fatu—and with them, their entire species—are running out of.

One could sense that Hildebrandt was in a great hurry. Not for the sake of his scientific legacy, but for "the girls."

Only Najin and Fatu are left to transmit the behavioral patterns of the northern white rhino to the potential baby northern whites. How to communicate, how to love, how to feed, play, fight, and rest. These traits are nothing short of a cultural template, a species' basic primeval blueprint. If we humans were raised by wolves, we would grow up to be wolf people. History is filled with examples of this sort.

"Not everything is down to the genes," Hildebrandt pointed out. "The environment—especially relationships—is hugely important as well. Only Najin and Fatu possess any real capacity to raise and to educate. Only they will be able to keep the species on track. I'm sure you're familiar with the story of Kaspar Hauser, who was raised in perfect isolation, inside a cage. Naturally enough, his behavior was that of an animal, and he ultimately committed suicide. The boy was human, yet nobody taught him to act like a human."

Hildebrandt made a brief pause. "Okay, so we know that Najin and Fatu are not exactly swapping tales, but a great deal of communication still passes between them. You probably know that with the orca whales, every group develops a dialect all its own. It would be stupid to suppose the northern white rhinos use the same language as the southern white rhinos. And that is why it's very important for the offspring to also inherit his or her parents' cultural know-how."

Hildebrandt kept repeating that a single baby would not be able to save the northern white rhino from extinction. However, it would certainly be an extremely important step in the right direction. "If we succeed, we will be able to demonstrate that science can help fix some of our most depressing problems. And that hope springs eternal!"

Incidentally, these are the precise sentiments that made us write this book.

We also believe hope springs eternal. If we didn't, we would find it very hard to hold on to our sanity.

Whether the northern white rhino is a subspecies of the white rhino, or an autonomous species is a question that scientists have not been able to agree upon. Hildebrandt, for instance, favors the term "ecotype."

"Rhinos are creatures shaped by evolution and they are fifty million years old," he said. "We have no right to wipe them from the planet. They play a very important role within their complex habitats. They always have. Like I said, if we remove them from the equation, this will sooner or later give rise to a new disease. As a scientist, my responsibility is to prevent that from happening."

Of all the BioRescue consortium members, Hildebrandt is the most insistent on transferring a northern white rhino embryo into a surrogate mother as soon as possible. In his opinion, completing this task is absolutely crucial to the project's success. "We cannot allow ourselves not to be in a great hurry."

Yet, as this book was being finished, the consortium still hadn't reached a consensus on the embryo-transfer timeline.

Hildebrandt, who says that the fight for the northern white rhino's survival has already turned his hair white, could not be clearer: "Several of my colleagues believe that I'm going too fast, that we still need more data. But the successes we've pulled off so far haven't been the result of waiting. No! They were the results of hurrying."

The project's beginnings were humble indeed.

"At the beginning, we had almost no funds," Hildebrandt recalled. "So, I had to develop a lot of the technology on a very tight budget. After we got 4.2 million euros from the German Federal Ministry of Education and Research, we were quickly able to take several huge steps. Mind you, we had been preparing for these breakthroughs for a long time. The simple fact is that we are under tremendous temporal pressure. This urgency must be our guiding light. It will be very hard for us to reach an agreement, but I am usually quite good at persuading people. So, I believe the eventual consensus will probably lean heavily toward my position, ha, ha."

The history of saving the northern white rhino is one of many ups and downs. "We have already performed the artificial fertilization and embryo transfer process with other large mammals, and we've had a fair deal of success. We first tried it with the northern whites at the beginning of 2000. Our attempt was a failure. So all we could do was to keep witnessing the species' dramatic decline," the German scientist turned our conversation back to the project of saving a functionally extinct species.

In a few ways, the project was initiated about twenty years ago, when a group of scientists started collecting the sperm of several males at the San Diego and Dvůr Králové zoos. The scientists were acting on pure instinct, without anything resembling a real plan.

The group also decided to study the reproductive behavior of northern white rhino females. Then the Dvůr Králové Zoo decided to relocate four of its northern white rhinos to the Ol Osaka Pejeta Conservancy.

"When they reached Kenya, we already knew Suni suffered from a tumor in his rectum, so he wouldn't live very long. He died in 2014, the same year as Angalifu, who lived at the San Diego Zoo. Angalifu's sperm proved of the highest quality from all of the samples collected. 2015 saw the passing of Nola at the San Diego Zoo, and Nabire at Dvůr Králové. Sudan, the last male, died in 2018. Only Najin and Fatu were left," Hildebrandt listed the casualties with horror in his voice. He was personally involved with most of the procedures.

By this point, the chronology of death he had just laid out was one the two of us knew almost by heart.

After Hildebrandt patented a device for non-invasive oocyte collection, the process of collecting the oocytes could begin in earnest. The gathered material was matured by Hildebrandt's colleague Cesare Galli from the Avantea laboratory in Italy. The oocytes were fertilized with the sperm of deceased males. After two years of hard work, Galli created the first hybrid embryo between the northern and the southern white rhino.

This was in 2018, while 2019 saw the creation of the first "thorough-bred" northern white rhino embryo.

On May 1 of that same year, the BioRescue project was awarded a generous grant by the German Federal Ministry of Education and Research. This meant a huge boost for the entire project.

Financial aid was also provided by Richard McLellan, a retired medical doctor from America. According to Hildebrandt, the million dollars donated by McLellan proved immensely helpful, since the ministry's grant was allocated only to the German team working on the project; there have so far been no guarantees that the grant will be renewed.

Over the past decade, the veterinarian from Berlin played a central role in the story of Najin, Fatu, and their potential offspring. In 2012, his team was given permission by the Dvůr Králové Zoo to examine the two females and determine the reason for their inability to conceive.

The answer was provided by an examination performed in 2014. Najin was diagnosed with tumors on her uterus and one of the ovaries. It was also determined she would be unable to successfully carry and deliver any offspring, since she was suffering from a severed Achilles tendon—probably caused by an overzealous suitor in the past.

Fatu, on the other hand, was incapable of conceiving on account of her endometriosis.

"The ovaries were their only active reproductive organs. We got in touch with Katsuhiko Hayashi from the Osaka University in Japan, who created the eggs by reprogramming the skin cells of mice. So that meant a small chance that the southern white rhinos might be saved after all," Hildebrandt recalled.

Additional momentum was provided when the German scientist collected $25,000 from the Pittsburgh Zoo & PPG Aquarium for his research work.

Hildebrandt decided the money would be best spent on saving the northern white rhino. "In the spring of 2015, I visited the San Diego Zoo to propose organizing a joint conference on northern white rhinos. They agreed, so I contributed half of the money from my own pocket, and the other half was provided by the San Diego Zoo."

The conference took place in December 2015 in Vienna, with the participation of twenty rhino experts from five continents. The event resulted in the formation of a strategic plan postulating two main approaches. One of them involved creating northern white embryos through the artificial fertilization of Najin and Fatu's eggs with the sperm of deceased males, then transferring the embryos into surrogate mothers from the northern white rhino's closest relative, the southern white rhino.

The other approach focused on creating embryos by fertilizing the eggs obtained with the help of stem-cell technology. Or, more precisely, by reprogramming northern white skin cells into egg cells.

It was agreed that the first approach would bring quicker results in the form of actual offspring, while the second approach would take much longer, but would ensure greater genetic diversity, and therefore contribute much to the species' long-term survival.

At the time when the strategic plan was formulated, none of the two options was feasible. Yet in 2012, the Americans demonstrated that the skin cells of the northern whites could be converted into induced pluripotent stem cells. "Which was one hell of an achievement!" Hildebrandt likes to point out.

At the start of 2016, the experts met in San Diego for another consultation on oocyte-collection techniques. "Then we parted ways with the San Diego Zoo," Hildebrandt regretfully reported. According to him, this was the moment when the collaboration between the project's European-Japanese and American branches began to collapse for no other reason than that the American scientists chose the path of competition.

It would be hard to find a worse moment for any sort of competitiveness.

Hildebrandt did his best to find layman's terms to explain how a lab-created embryo will be transferred into a living animal.

"Embryo transfer is the simplest available technology for the artificial fertilization of human beings. We also use the procedure with horses. The process gets a bit harder with cows. With sheep, we have to do it endo-scopically and with the northern white rhinos, the protocol is also pretty demanding, given the species' size and the morphology of the cervix. The procedure can only be completed with the help of general anesthesia."

In collaboration with his colleagues, Hildebrandt developed a tech-nique involving a modified ultrasonically guided catheter with a needle at its end. The device aims to transfer the embryo into the uterus through the rectal wall.

By the end of 2021, the procedure had already been performed on southern white rhinos eight times. In two cases, the embryos gave signs that they might develop further, yet they eventually did not survive.

This was because oocyte collection and embryo transfers were performed on females from European zoos; all of them suffered from reproductive problems. The team had not been given permission to work with females whose reproductive organs were functioning soundly. So, the project's leaders were overjoyed when Kenya Wildlife Service gave them access to females who had already successfully delivered offspring.

At the Ol Pejeta Conservancy, a pair of southern white females were moved to the pen holding the castrated southern white male named Owuan. Several long months were spent studying their reproductive cycles.

"Since there's still insufficient evidence that our new technique works, the Dvůr Králové zoo proposed we first transfer a southern white embryo to a surrogate mother and then if the procedure proves successful, we can try with a northern white embryo. We've already obtained per-mission. We also have first-class southern white embryos," Hildebrandt described some of the more recent developments in early 2022.

Thirty days after the transfer of the southern white embryo is com-pleted, the Ol Pejeta female will be given an ultrasound scan. If she passes, the pregnancy will be terminated, and another female will begin her preparations for the transfer of a northern white embryo.

With all this about to happen, Hildebrandt is also preparing for the potential use of assisted-reproduction technologies on different types of rhinos and other large mammals.

A good example would be his close collaboration with his Chinese colleagues, who are working on the giant panda reproduction program. Pandas can normally reproduce in captivity, though some Chinese breeding centers are home to several animals who do not seem to possess the urge. Since these animals come from wild populations, their genes are very important for the species' overall genetic diversity.[1]

"Regarding the pandas, the Chinese asked for our help with oocyte collection, artificial fertilization, embryo creation, and embryo transfers, but the Covid-19 pandemic significantly slowed our collaboration," the German scientist explained.

The techniques Hildebrandt had developed could eventually be used on other critically endangered species, like the okapis.

He has also done a lot of work with Sumatran rhinos. "We collected more than forty oocytes, but the males' sperm was of very low quality, and we were unable to obtain enough quality sperm from Indonesia. The other problem with Sumatran rhinos is their expected lifespan, which is shorter than with other types of rhinos. At most, they reach thirty years of age. They are also highly susceptible to different kinds of tumors: on the uterus, on the kidneys, on their faces. . . ."

Naturally, views on saving a functionally extinct species vary.

Over the phone, John Lukas—the International Rhino Foundation president—informed us that his attitude towards Hildebrandt's project was at least somewhat ambiguous. In total, the tissues of twelve animals and the sperm of five males had been stored. Lukas believes the available gene pool will prove insufficient to create a new northern white rhino population.

"This is a scientific project involving lots of people. They're developing important technologies for the future, but the project is geographically limited. Fifteen years ago, we had a chance of protecting the northern white rhinos at the Garamba National Park in the Democratic

Republic of Congo, and thus preserving them in their natural habitat," Lukas reminded us of a wasted opportunity.

"Back then, we were unable to arrange sufficient levels of collaboration with the various zoos, so that all the Garamba rhinos could be relocated somewhere safe. The most sound and realistic option was Ol Pejeta, where everything was ready for the relocation of five Garamba's northern whites. Unfortunately, that didn't happen," the legendary American conservationist deplored that attempt's outcome.

Should the project of saving the northern white rhino prove successful, Lukas believes it would offer the world hope that critically endangered species might be able to be saved after all.

He identified one of the key problems faced by African conservationists: "We have to ask ourselves whether it makes sense to insist on leaving the endangered animals in their natural environment, where they are at their most vulnerable and exposed. The case of the southern white rhinos and the black rhinos demonstrated that relocations work. We could also easily move the animals to Uganda, Tanzania, or to the Garamba itself. Yet we'd be faced with brutal nationalism and other aspects of the local political context. Which are usually not at all favorable to the conservation of endangered species."

John Lukas also mentioned that many members of the scientific community wondered why BioRescue would waste "all that time and money" on saving the northern white rhino.

"The fact is that only two females are still alive. It is also true that only seventy-five Javan rhinos and eighty Sumatran ones are left in the entire world. Both ancient species still live in their natural environments and both would certainly need and deserve a lot more attention and funds than they are getting," was Lukas's realistic appraisal of the situation.

"My own efforts regarding saving the rhinos are thus mostly focused on Asia. We've been rather successful in India and Nepal, where the governments are actively involved in saving the Indian rhinos—through financing, providing security, and legal prosecution of poachers. The local rhinos have benefitted enormously. The Indian rhinos living in India and

Nepal are certainly a success story, mostly because there was a strong political will to protect them. Even the army and the police are taking part in the rhinos' protection. The whole approach is holistic. The authorities have done a great deal to minimize the conflict between humans and wild animals. Yet even so, the rhinos in the mentioned countries remain highly vulnerable," Lukas explained.

The political and environmental context that confronts the African rhinos could hardly be more different.

"The greatest difference is the physical environment. In Africa, the rhinos—especially the white ones—can be spotted from a long way off. They live out in the open, which means they are more exposed. They are also substantially larger. Their Asian counterparts, on the other hand, mostly live in the forest, so they are more hidden. We can only find them with the help of camera traps or frequent patrols. This is part of the reason why it is so hard to determine how many of them are still left in the natural environment," John Lukas reported.

Having spent three decades operating in the Democratic Republic of Congo, Lukas was able to observe the northern white rhinos while they were still living in the wild.

The Javan rhinos only live in one place, which makes them a lot easier to monitor. They are, for example, very well protected within the Ujung Kulon National Park in Indonesia, but not outside of it. The steep rise in human population and the shrinking of the forests pose a significantly greater threat to them than the poachers. Another threat is the highly active Krakatoa volcano.

"This is why we want to set up a so-called substitute population at another location somewhere in the wilds. We are currently negotiating with the Indonesian government, which needs to show some political will to relocate the animals. Only then will we be able to start searching for the actual location. Indonesia is a rather decentralized country and Java acts very possessively towards its animals, though that's understandable," Lukas described the realities of saving rhinos at the other end of the world.

The situation is similar in Sumatra, where the rhinos are vulnerable not only to climate change, poachers, and the increase in human population, but also to natural disasters. A few rhinos perished during the December 2004 earthquake and tsunami. Sumatra, however, is substantially more wooded than Java and—despite all the problems—the living conditions of its rhinos are excellent. For this reason, the International Rhino Foundation wanted to move a part of the Javan rhinos to Sumatra, but it was faced with strong resistance from the Indonesian authorities.

Lukas spent many years working at various zoos, which he believes to be the sole remaining option for saving at least some of the critically endangered species.

"There's a huge difference between a zoo and a reserve like Ol Pejeta," he pointed out. "At Ol Pejeta, the animals can roam freely. Fences and rangers provide protection from poachers. This is one of the shapes conservation can take. You can find several similar cases in the South African Republic, where the southern white rhinos and the black rhinos live in fenced-off and heavily protected parks. They're allowed to live in a semblance of their natural environment, which could never happen at a zoo. On the other hand, the endangered populations residing at the zoos are hugely important for preserving a sufficiently wide genetic pool."

In Lukas's opinion, saving the northern white rhino will doubtlessly involve the help of stem-cell technologies. The same goes for saving many other species.

"I am convinced the development of stem cell technology will prove crucial for the global rescue of endangered species," Lukas served up some much-needed optimism. "With the help of said technology, we will perhaps be able to do things we can currently not even begin to imagine."

Raoul du Toit's views are quite similar. He believes that the functional extinction of the northern white rhino was not solely caused by the conditions in Africa, and that the role of zoos calls for a critical revaluation.

Even thirty years ago, du Toit favored crossbreeding northern whites with the southern whites. In his view, the northern white rhino is really a subspecies of the southern white rhino. The Zimbabwean zoologist remains convinced that had his recommendations been heeded, the crossbreeding would have quickly boosted the overall genetic diversity, and the northern white rhino might have been saved.

"This would have been the only way to ensure both subspecies' evolutionary continuity, especially in the areas not traditionally inhabited by the white rhinos. For instance, in Kenya. Instead, we opted to separate the two subspecies—which, at least in my view, was a mistake. It was the general consensus at the time, but it caused irreparable damage to the species. What we did is to create a bottleneck. Trying to save a subspecies, we unfortunately put the entire species in danger," du Toit criticized the first attempts at saving the northern white rhino.

Du Toit also has his reservations about saving the northern white rhino through advanced assisted-reproduction technologies. Yet he doesn't see any other available solutions. The continuation—even, so to speak, *resurrection*—of the species is now entirely in the hands of science.

Yet even so, du Toit believes it will take nothing short of a miracle.

"Far too much faith has been placed in our technology, and far too little effort made to pursue simple solutions—like the crossbreeding initiative we discussed earlier," he recalled the decisive years for the (sub)species' survival. "Such high hopes were riding on artificial fertilization at the zoos, but these programs proved completely ineffective and even caused great harm."

For Raoul du Toit, years of failed attempts at artificial fertilization came as no surprise since he had long familiarized himself with the exceptional complexities of natural rhino reproduction and with the highly demanding nature of artificial-fertilization procedures.

"It is not my place to comment on the developments at Ol Pejeta," he smiled politely. "I hope for the best and wish them all the luck in the world. But unfortunately, it seems we have already missed out on our key opportunity for saving a subspecies."

In his youth, du Toit worked in Zimbabwean water management. This was how he came into contact with conservation and rhinos. He then spent seven years working in national parks. Even then, in the 1980s and the 1990s, the rhino's endangerment was one of the main topics of discussion.

Du Toit became completely obsessed with trying to protect the (sub) species.

To him, the rhino's fate epitomized all of the key ingredients of conservation: ensuring the animals' welfare, taking on poachers, activating the political system, seeking out sources of financial support, international cooperation, various guerrilla activities, the economy, property laws, climate change, the consequences of colonialism, wars, geostrategic tectonics, the conflict between the world of humans and the world of wild animals, cooperation with the local communities, trade, the legal system, security, agriculture, natural resources, and especially water.

Even then, du Toit knew that the story of the vanishing rhinos was a story that involves each and every one of us.

"When we're saving the rhinos, we're really saving entire ecosystems. We are rescuing biodiversity. We are helping the local communities. We are bolstering the economy and promoting general harmony. The rhinos should be placed at the very heart of our efforts. To me, the fight for these animals represents the frontline of all conservation," du Toit summed up his life's calling.

Thomas Hildebrandt, the tireless BioRescue leader, is of course aware of the reservations concerning his project of saving the northern white rhinos. He understands the substance of these reservations, yet he firmly disagrees.

"Some people believe that we are playing God," he told us. "This is why we need to be very open about what we're actually doing. We really need to be careful to avoid the fate of the scientists who cloned Dolly the sheep. The public ultimately perceived them as monsters who had gone against the laws of nature."

Hildebrandt believes that when faced with a task as important as saving the northern white rhino, the scientific community should strive to unite. The German scientist reiterates his appeal again and again. Yet so far, it hasn't come to pass.

There are very few signs that it still might. The countless particular interests powered by corporate and personal egos are apparently stronger than the imperative of rescuing a practically extinct species.

"It's traumatic, being so powerless while an entire species vanishes," Hildebrandt related. "In this context, I was quite riled up by the billionaires' race to 'conquer' space. Those rich idiots have no qualms about throwing unconscionable amounts of money on that, while our own planet is badly damaged and urgently needs help. I feel that the money would have been a lot better spent on saving the critically endangered animal species."

To put things in perspective, a single space sortie by Elon Musk or Jeff Bezos costs more than the entire projected budget for saving the northern white rhino.

In his fight for critically endangered species, Thomas Hildebrandt keeps facing the dark and so-often dominant part of human nature. So, we had to ask: did he, like countless war reporters, begin to lose his faith in humanity?

"Well," he replied, "as a species, humans are similarly intelligent as elephants. Yet they are also much more aggressive. Humans are highly inventive and not very kind. They are capable of creating the most beautiful things. Just think of Michelangelo, da Vinci, and Tchaikovsky! But they also create unspeakable horrors like concentration camps. When pursuing their own selfish goals, they can lay waste to entire regions, paying no mind to the other species. Let's put it this way: humans will quickly have to discover some very hidden depths in their nature, or they will be gone."

Or we will be gone.

CHAPTER FIVE

Creating Life

A GROUP OF TOURISTS RIDING SLENDER BROWN HORSES WERE LET inside Najin and Fatu's pen inside the Ol Pejeta Conservancy. The two of us exchanged a startled glance. We were surprised that anyone besides caretakers, rangers, veterinarians, and journalists was allowed to enter the home of the last two.

Yet we quickly realized that, during our first visit to Ol Pejeta, that the heavily guarded pen was unvisited solely on account of the Covid-19 lockdown. On February 19, 2022, tourists were allowed to return to the Conservancy, which had been hit rather hard by the pandemic-induced fallout.

Horseback visits of the last two northern white rhinos are one of the many tourist activities offered at Ol Pejeta. The cost is the rough equivalent of a daily ticket at a Swiss ski resort. The experience, however, is unforgettable: getting close to the last two members of a species up to a horn's length. This is, of course, as close as you can get.

While taking in the unusual visitors as the morning began to swelter, we were surprised by another observation: that Najin and Fatu, who normally get enormously upset at every bird emitting a slightly-too-loud chirp, at every tiny mouse darting under their feet and sometimes even at a locust springing up somewhere nearby, now seemed to be not the slightest bit disturbed by the horses.

In fact, it was quite the contrary. The two rhinos shifted their attention from their constant fixation on grazing to the graceful equine guests.

Their two horned heads lifted up and sniffed the air wafting over from the newcomers.

The horses equally didn't seem to mind. They languidly stood in place, driving off flies with their tails. Najin and Fatu approached to a few meters' length. Their snouts wanted to edge even closer; it was as if the two rhinos wished to establish physical contact. The rapt chatter of the riders suddenly stopped. The tourists were clearly frightened by the close proximity of a pair of horned behemoths, even though the situation was carefully monitored by both the caretaker Zachary Mutai and the leader of the mounted expedition.

The riders opted for a strategic retreat, though the horses were in no hurry to obey. Like Najin and Fatu, they, too, felt like hanging out.

Why were the rhinos usually mortally frightened by the tiniest of field creatures, but showed no fear at the sight of horses? How come the horses had no fear of the rhinos? Why did it feel like the animals were actually trying to communicate, and even establish a rapport?

Next to zebras and tapirs, horses are the rhinos' closest surviving relatives (all of them are odd-toed ungulates). As the closest relatives among domesticated animals, horses are to play an important part in saving the northern white rhino from extinction. Because he adapted the method of artificial fertilization and embryo creation used on elite sport horses, the person most responsible for this was Italian embryologist Cesare Galli.

Cesare Galli screwed off the lid of a liquid nitrogen cannister and pulled out a sample labelled '*Fatu NWR x Suni NWR, 9. 4. 21.*' Awestruck, we were afforded a glimpse of an invaluable rarity: northern white rhino embryos, created in March and April 2021 at the Avantea lab in the Italian city of Cremona.

By the autumn of 2022, the Avantea biotech company, which specializes in advanced animal reproduction technologies, managed to create twenty-two northern white rhino embryos. In 2019, the sixty-one-year-old Cesare Galli became the globe's first scientist to pull off such a feat.

To be clear: we are talking about blastocyst stage embryos mature enough to be transferred into the uterus, where they are able to develop further—not about one of the earlier stages, when the embryos are still fragmented and prone to falling apart. This was precisely what happened to the scientific team which boasted of creating a northern white rhino embryo a year before Avantea, Galli pointed out. When, in reality, all the above-mentioned team did was to create a few cells.

Even though the cofounder and managing director of the Italian company created the first two northern white rhino embryos a mere three weeks after the oocytes were collected from Najin and Fatu at Ol Pejeta, the triumph was the result of five long years of planning and honing the procedures. The goal demanded the cooperation of all of the key members of the BioRescue project. This is why Galli is profoundly proud of the achievement, even though his face is normally hard to read for emotion.

He is also proud of having created the world's first cloned horse. Prometea came to be in 2003, and Galli discusses her in much more sentimental terms than he does the northern white rhino embryos.[1]

"Prometea was born on May 28 in 2003. She is still with us. I remember it so well! It was around midnight when I received the call that the alarm on the vulva of Prometea's mother—who is also her twin sister—had gone off. The vulva was beginning to dilate. When I got to the stables around half past one, the foal was already halfway out. I almost fainted. I couldn't believe we'd actually done it! When we created the first northern white rhino embryos, I didn't faint, but mostly because I was sitting down," Galli laughed.

At twenty years of age, Prometea, whose asymmetric face lends her the appearance of having been painted by Pablo Picasso, is healthy and very lively to boot. She lives with her mother at the stables next to a building hosting Avantea's labs.

After Prometea, Galli's intent was to start cloning genetic copies of elite sport horses.

But the limitations of cloning—high embryo mortality rates during pregnancy, high mortality rates of offspring soon after birth, or numerous medical difficulties later in life—put him off from taking his experiments further. Another part of the reason was the rather negative general attitude towards cloning in Europe.[2]

And so, Galli redirected his efforts toward creating horse embryos through assisted reproduction.

From 2000 on, the Italian company located amid the fields near Cremona has been creating elite sport horse embryos. The key was the so-called ICSI technique, or the direct injection of a sperm cell into the cytoplasm of an egg cell. By 2006, the company had already sufficiently tested out the technique to start commercializing it.

In this field, Avantea's expertise is second to none. "From the marketing point of view, our core activity is assisted reproduction of horses," Galli reported. "These animals do not respond well to standard *in-vitro fertilization* procedures. So, we must use technology developed to cure infertility in humans. This is where the ICSI method comes in. By now, it has been developed to a very high standard, and there is precious few of us who know how to perform it."[3]

In 2001, Cesare Galli became the first scientist to create a horse embryo with the help of the ICSI method.

The technique itself is one of the most effective methods at our disposal for curing infertility in human males when too few sperm cells are present—or when the sperm cells are not mobile enough, or not properly shaped. ICSI is thus the most widely used method of artificial fertilization in humans. It proved ineffective with horses until Galli came up with his adjustments.

"While we use a sharp-tipped pipette when trying to insert human sperm into a human egg, we learned it was crucial to use a blunt tip on the animals, since the sharp tip damages the egg," Galli described one of the discoveries.

The man who also used to serve as a professor of animal reproduction at the University of Bologna patiently explained to us that the assisted reproduction of sport horses was "a method of breeding males and females of high genetic quality." If a female is used for racing, she is usually prevented from breeding in her younger and more fertile years. Later, her fertility is much diminished, so medical assistance is needed with fertilizing the eggs.

"A rather similar principle applies to the males," Galli shook his head in commiseration. "Those who've had a successful career—well, their sperm is either the quality of an old horse's sperm, or they become infertile altogether. Mind you, many of the stallions are unable to reproduce in the natural way on account of being castrated."

Assisted reproduction also has a commercial advantage. A very small quantity of sperm enables the production of a very large number of embryos and foals. "If a male died in an accident or from a disease, and very little of his sperm is available, the method helps you fertilize a great many females nonetheless," Galli clarified.

At the moment, ICSI remains the only successful method of artificially fertilizing horses, but it consistently delivers results, just as it does with humans.

We were able to observe the procedure on a computer screen behind the Cremona laboratory's glass wall. While carefully preparing the egg cells and the defrosted horse sperm, the diligent assistants Silvia, Gabriella, and Paola were monitored by Galli's wife, Giovanna Lazzari, the co-founder and scientific director of Avantea.

The assistant in charge of the artificial fertilization used a pipette to insert the sperm into an egg cell. Pushing the fertilized egg aside, she wasted no time in pulling up the next one. This was the exact same procedure used by the Italian company to create northern white rhino embryos. The fascinating scene of routine life creation was presided over by Jesus hanging from a cross on the wall.

"Women are more suited to this particular task. They are more patient and conscientious, but it's hard to find suitable employees," Galli related.

In 2020 alone, Avantea created five thousand horse embryos. The company's clients are the owners of elite sport horses from Europe— especially Eastern Europe—and from the Middle East.

"The usual procedure is to transfer the embryos into young females brought to us by the owners," Galli explained. "The females are then immediately taken back home, since the owners wish to have the foals born in their respective countries. We are also often only sent the eggs, so we can fertilize them in our lab and ship back the embryos."

"Is this big business?" Galli repeated our question, standing next to the afternoon's crop of fresh frozen life. "It depends on what you mean by big business. It is certainly a niche and we are the only ones who are able to do it on a professional level."

In the case of the Italian scientist with over thirty years of experience in embryology and assisted reproduction, there was certainly no need for false modesty. It was Galli's knowledge of the secrets and potential pitfalls of horse reproduction that opened the gate to his inclusion in the BioRescue project. Just to reiterate—among all the domesticated animals, horses are the rhinos' closest genetic relatives.

Wasting no time, Avantea adjusted its protocols of oocyte collection and maturation, egg-cell fertilization and embryo creation to suit the reproductive needs of northern white rhinos. The company also began testing out a technique of transrectal transfer of embryos into horses. When the time comes, the surrogate southern white rhino mothers will receive the northern white rhino embryos through their rectums, since it is not possible to do it through their vaginas.

Avantea has some of its own financial capital invested in the BioRescue project. Yet the sum is considered negligible given the boost in global recognition and all the new clients attracted by the firm's participation in one of the best-known and most inspiring conservation projects of our time.

Creating northern white rhino embryos was a great triumph, and also a very emotional experience, Galli nodded. "But then you want more, like an addict. So now we need our next fix—a baby northern white rhino!"

Cesare Galli grew up on a dairy farm where chickens, turkeys, ducks, goats, and sheep were also raised.

"As a child, I realized the key to breeding animals was to understand their reproduction. That is why I later decided to study veterinary medicine, focusing on larger animals, like cows," the Italian scientist reminisced in his modestly furnished office, where he was surrounded with photos of cows and horses and a single image of a rhino in a bubble.

During his studies at the University of Milan, Galli determined he was most interested in developing new reproduction techniques for creating "better" animals. He was never tempted by a veterinary career. What he learned at the university; he was able to test out on his home farm as he went along.

During his last year of studies, Galli started doing lab work with cow oocytes. He got his doctorate based on the strength of his thesis on superovulation in cows. His doctorate supervisor put him in touch with Cambridge University in Great Britain, where Galli promptly moved after completing his studies in Milan.

His post-doc period at Cambridge was marked by research into various reproductive biotechnologies, which involved cloning. Galli was focusing on horses, pigs, and cattle, and on creating animal models for biomedical research.

Between 1986 and 1988, he also found time to work for the Animal Biotechnology Cambridge start-up company, whose goal was to start creating cattle embryos on an industrial scale. In 1987, they managed to produce their first calf.

Galli found he didn't really feel at his best working for the start-up company. He especially resented being unable to publish the results of his research. He had, after all, pledged his young and idealistic life to science. So, he accepted the offer of a research post at the Babraham Institute, one of Cambridge University's partner organizations.

After two years of working there, Galli spent the next three years at the Biotechnology and Biological Sciences Research Council institute. He began perfecting his understanding of the somatic cell nuclear transfer method (SCNT), the most widely used technique for cloning higher organisms. SCNT eventually gave birth to the world's first cloned mammal, Dolly the sheep.[4]

At his new post, Galli spent some time collaborating with Martin Evans, who was later awarded the Nobel Prize for his discovery of embryonic stem cells and subsequent research.[5]

After a series of failed experiments with embryonic stem cells, however, the institute opted to focus on cloning—a field awarded wider financial support at the time. The institute's aim was to develop a procedure for cloning "elite animals" *en masse*. The idea was to collect embryonic stem cells from, say, lions and start cloning them into infinity.

When this venture failed as well, the institute decided to bank on cloning through the use of somatic cells (a somatic cell is every single cell in the body, except for the reproductive ones). During that same period near the end of the eighties, Ian Wilmut and Keith Campbell, who later cloned Dolly, also experimented with the SCNT cloning technique at The Roslin Institute.

"We exchanged information and even collaborated in parts, since it was the same institute, only at a different location," Galli nodded. By the end of 1989, he himself was already cloning animals using the same technique that later resulted in Dolly.

Yet at the time, the entire field was still in its embryonic phase. Galli's efforts resulted "only" in several animal pregnancies, yet not a single live-born offspring. Before he could reach that goal, his post-doctoral grant ran out.

While doing research at the British institute, Galli collaborated with Giovanna Lazzari, an expert in assisted reproduction whom he later married.

The couple decided to return to Italy. Part of the reason was that the institute stopped doing experiments with large animals and began using mice. The Gallis accepted an offer from the Italian Breeders Association to set up a similar institute in Italy, focusing on the mass creation of cattle embryos and their transfer into cow females.

In 1991, they moved to Cremona and set up a laboratory at what used to be a cattle farm. "The company consisted of me, Giovanna, a technician, and a veterinarian. Using the classic artificial fertilization method, we set out to create cattle embryos. We got our oocytes from the ovaries of females in slaughterhouses. Yet our business plan didn't hold up. Our added value proved insufficient, given the price and complexity of embryo-creation technologies. So, we decided to halt our activities," Cesare Galli related, matter-of-factly.

Embryo creation via artificial fertilization became an established practice in livestock breeding later on, when it was combined with the collection of immature egg cells from live animals. The procedure was developed by a Dutch team of scientists from Utrecht University.

"According to this method, only the most precious animals can become oocyte donors. Their preciousness is determined by the quality of their meat or their capacity for milk production. Through artificial fertilization of their eggs, you can start creating embryos *en masse*," explained Galli, who also serves as the president of the International Embryo Technology Society.

Over the past several years, more calf embryos were created through artificial fertilization than in the natural fashion: three quarters of the total.

When questioned whether Avantea's technologies had provided a huge boost to the livestock industry, Galli responded: "A little over thirty years ago, we were only starting to use these technologies. Today they are the prevalent method all over the cattle breeding industry. Much the same goes for horses. Yes, the technology we've developed had a huge impact. It was also crucial for the development of the oocyte-collection techniques. But you see, being first doesn't always mean being the most successful."

In 1994, Galli went back to cloning. In 1999, his scientific ship came in when he became the first to clone a baby bull. He named him Galileo after the Italian physicist, astronomer, and engineer, who proved the Earth revolved around the sun and spent the last years of his life under house arrest as punishment.

Galli seemed to be facing a similar fate. When he brought Galileo to a livestock exhibition in Cremona, he was arrested by the police. He spent the night in prison, while the bull was confiscated and only returned three months later.

After Dolly, the Italian government instituted a ban on all further cloning of animals and humans. Galli was charged and tried; the judge ruled in his favor.

"The whole thing was almost funny," Galli sighed. "At the time, the minister of health was Rosy Bindi, an ardent catholic who didn't see much difference between cloning humans and cloning animals, so she banned both."[6]

Another calamity struck in the year of Galileo's birth. Mad cow disease broke out again in Great Britain. The livestock trade collapsed overnight.

"Since the breeders were so afraid of having to cull entire herds, hardly any animals were slaughtered for a long time and we were beginning to run out of material for oocyte collection," Galli related. "We decided to start using our artificial fertilization techniques to create embryos of pigs and horses, who were unaffected by the disease. We also considered branching out into embryos of cats and dogs."

In the end, the creation of sport horse embryos seemed to be the soundest option, though the company also increased its involvement with genetically modified pigs. In this field as well, Galli was quite the trailblazer. In 2006 he created Italy's first cloned pigs. Two years later, he produced the country's first genetically modified pigs.[7]

"We came into contact with rhinos through the technology we developed for horses," Galli explained the switch to saving wild animals, "though pigs were also involved."

The Avantea lab creates genetically modified pigs for biomedical research. When Galli had to deal with an infertile male, his colleagues advised him to contact Dr. Thomas Hildebrandt, one of the pioneers of assisted reproduction in large mammals.

"He was said to be able to secure the pig's sperm with the help of electroejaculation," Galli recalled. "Hildebrandt and his team came over, got the sperm and used it to fertilize the females. We got what we wanted!"

This is how Cesare Galli described the first encounter of the two fathers of the BioRescue project.

Then it was Hildebrandt's turn to ask for his colleague's help in creating northern white rhino embryos.

"We started out with Sumatran rhinos," Galli nodded. "At the time, we didn't have permission to work with the northern whites, since Kenya wouldn't allow Najin and Fatu's oocytes to leave the country. And so, between 2013 and 2015, I made a number of trips to Borneo."

The Sumatran rhino is smaller than the African varieties. It is about the size of a horse. "At the local zoo they had a pair of very old females we used for oocyte collection. They also had a single male with awful sperm quality, while no frozen sperm was available. Our lab was unable to create embryos from the material on hand, but we still learned a lot."

The company then started working with southern white rhinos. Since they were located in European zoos instead of in Africa or Borneo, the process of oocyte collection was logistically much less challenging. 2015 brought the first opportunity to collect oocytes from a northern white rhino female. That was the year Nabire, one of the last four remaining northern white rhino females, passed away. She lived in the Dvůr Králové Zoo in the Czech Republic.

"One morning her left ovary was delivered to us. We managed to collect, mature, and fertilize the oocytes, yet nothing came of it," Galli revisited a period of great disappointment. He quickly added that this was before the company's protocol for the assisted reproduction of horses had been adjusted for northern white rhinos. The adjustments were developed between 2015 and 2019.

In 2018, Galli created the first two hybrid rhino embryos by fertilizing a southern white rhino egg cell with the sperm of a deceased northern white male. The southern white egg was donated by a southern white female appropriately named Hope.

That same year, some good news arrived from Kenya: the government allowed Thomas Hildebrandt to collect the oocytes of the last two northern white rhino females at the Ol Pejeta Conservancy. Galli was sent the first batch of their oocytes in 2019.

He was finally able to get to work on some proper material.

In Galli's opinion, success isn't far off—the birth of a baby northern white rhino. "Only the techniques need to be made more consistent, a little more robust," he cautioned. "The success-rates have to be driven a little higher."

It took five years of research and preparation to create the first northern white embryo. "It was impossible to progress any faster. We couldn't skip a single step. There is precious little information available about the northern white rhinos and their physiology. Everything we do is new, meaning we can't draw on existing scientific findings," Galli summed up the project's pioneering spirit.

One of the initial obstacles to creating the embryos was the scarcity of available northern white rhino oocytes. Najin and Fatu's immature eggs were collected every few months—until the fall of 2021, when Najin's were stopped being collected altogether. Up to that point, every single embryo had been created from Fatu's eggs.

Another limitation placed on the project is the quality of the available sperm. The frozen sperm of four different males is on hand from Saut, Suni, Sudan, Angalifu, and Dinka.

There are also still many gaps in our understanding of the rhino sexual cycle. In Galli's view, this is partly because the Ol Pejeta Conservancy failed to do its homework. "Though it wouldn't be easy," Galli added. "They would have had to perform an ultrasound scan every single day over a month. Each time they would have to put Najin and Fatu to sleep, and the rhinos don't take sedation very well."

The sexual cycle could also be studied on southern white females in various zoos. "Several people have also raised orphaned rhinos," Galli ventured. "Though working with them would bring its own problems."

The success rate for embryo creation is still low. From 2019 to when we finished writing this book, oocytes were collected from Najin and Fatu ten times, netting the research team 140 oocytes.

As already mentioned, a total of twenty-two embryos were created by October 2022. "But you see, even with horses, only twenty percent of collected oocytes are turned into embryos strong enough to be transferred into the mares," Galli admitted. "With rhinos, the rate is about ten percent."

The embryos of northern white rhinos are stored in liquid nitrogen in Cremona and Berlin. In all cases, the role of Eve was played by Fatu. The role of Adam was filled by a pair of deceased males, Suni and Angalifu.

In their deep-frozen state, which halts all biological processes, the embryos are able to endure for decades or even centuries. Yet the scientists gathered within the BioRescue project have no intention of waiting that long. "We know the embryos are alive," Cesare Galli confided over the telephone at the start of 2022.

In his opinion, one of the key problems will be identifying whether the surrogate mother is ready to receive the embryo. "Our success depends on how many females we will be able to reach. I believe our basic purpose is achievable, though the logistics remain the main obstacle. We will have to perform initial tests and transfer several embryos," the author of over 180 contributions to international scientific journals explained.

We last talked to Galli on the day when Thomas Hildebrandt transferred a southern-white embryo into a surrogate mother at Ol Pejeta. It was the winter of 2022.

"We've decided to postpone the first northern white rhino embryo transfer until the success of the first southern white rhino embryo pregnancy," Galli related. He was very sorry that a trip to the United States and numerous pandemic limitations prevented him from joining Hildebrandt in Kenya.

Covid-19 slowed down the project in several ways. Since European zoos were closed, Hildebrandt was unable to collect southern white rhino oocytes, so Galli was unable to create the new embryos needed for testing the process at Ol Pejeta.

By the end of January 2022, only two southern white rhino embryos were still available. Galli seemed quite perturbed by not having access to the material required for the creation of new embryos.

Four weeks later, there was some additional bad news. The southern white rhino embryo that Hildebrandt had transferred into a surrogate mother did not survive.

On the question of how long the pregnancy of the first southern white rhino female should last before a northern white rhino embryo is transferred into another female, Galli's views differ from Hildebrandt's. The German scientist would prefer to transfer the northern white rhino embryo as soon as possible. Galli, on the other hand, believes the first pregnancy should ideally be carried to term until delivery. A successful birth would mean the experiment was 100 percent successful.

"The first breakthrough at Ol Pejeta will be if the embryo takes. So far—by February 2023—we haven't been able to manage that. It will be the most important step toward the eventual birth of the next northern white rhino. Once we have sound results with the southern white rhino embryos, we'll be able to deploy the northern white rhino embryos with confidence. Until then, I believe we would be ill-advised to do so. There are still far too many unknowns," the experienced embryologist cautioned.

Yet he, too, remained hopeful that the first pregnancy involving a northern white embryo would not take too long. "When people ask me: 'When can we expect the first northern white rhino infant?,' I usually say: 'In five years' time.' But I've been saying that for the last two years. It's next to impossible to make accurate predictions, what with the pandemic and the rising costs. But if we're able to do it, I don't see why we shouldn't! And I firmly believe we are eventually bound to succeed."

Galli realizes that the genetic material of two males (Suni and Angalifu) and a single female (Fatu) will be insufficient for the creation of a self-sustaining population. "But if we make embryos successfully, we could have enough of them in four years' time. If we achieved a fifty percent success rate of pregnancies, like we already have with horses, we would produce eight animals over four years. With the current technology, we could utilize the sperm of three or four deceased northern white rhino males. All right, so we couldn't exactly populate Kenya with eight rhinos, but it would be a good start."

For the Italian scientist, the necessity of saving the northern white rhino from extinction was never in doubt.

"I am not a wildlife expert. I am a veterinarian who mostly works with livestock," he smiled. "But I do know that if you take one species out of an ecosystem, the entire balance can get disrupted. The rhinos are one of the species which help maintain the natural balance. They are also an iconic species. Every child knows what a rhino is. If they were to go extinct, it would be because of us. True, the rhinos are not among the most fertile animals, but their current grim situation is our doing. This is just one of the reasons we can't allow them to vanish forever."

Galli also reminded us that the southern whites had been facing a similar situation as the northern whites. However, after the South African Republic acted to protect them, their number rose to above 20,000 again. "The northern white rhinos' additional misfortune was to populate areas that saw especially many wars," Galli added.

How does the wider scientific community regard the project of saving the northern white rhino, we asked the Italian scientist. "Well, those in similar lines of work as us realize very well what we're trying to do. They respect our work. They understand the magnitude of the problems we're facing. But ours is a niche field. It is quite hard to understand all the technical details. So no: thus far, we haven't seen a lot of interest or response from the wider scientific community. The media were a different case. I do, however, agree that a successful pregnancy would bring a tremendous boost to our cause."

Galli awaits that prospect with all the hope he can muster. But in the meantime, he has a company to run.

On average, he is able to allocate about a fifth of his working time to saving the northern white rhino. "It is certainly a challenge, but I am glad to be a part of it all. It normally only gets overwhelming during those final stretches, when the embryos are created in the lab after the oocytes had been collected. In the end, it all comes down to proper workplace organization."

Besides its bread-and-butter commercial programs of livestock assisted reproduction, Avantea also performs the procedure for various national and international research projects. In addition, the company is very active in the field of biotechnology research. It develops animal models for xenotransplantation (transplanting organ tissue from one species to another, especially from animals to humans) as well as animal models for studying human diseases and testing out medicines. The company performs advanced stem-cell research and develops alternative toxicology tests for various industries.

Only two days before our conversation in January 2022, Galli returned from the United States. So, we naturally assumed he had taken part in the first transplantation of a pig heart to a human being. We knew that Avantea also made pig models that were similarly genetically altered as the ones used in the United States.

However, Galli had nothing to do with the seminal medical achievement. "In Europe, there's a lot of fear of both cloning and genetical modification. What they just pulled off in America is what hardly anyone here would even dare think about! After the U.S. pig heart transplantation proved successful, I received several calls from people who want Avantea to start breeding these sorts of pigs. But by now, it's already too late for us to become the first."[8]

Avantea clones pigs and uses them in medical research. Most of the cloned pigs are sent to other companies for experimentation. Each year Avantea also clones a few horses. It utilizes the CRISPR-Cas9 method of gene editing to create cattle intended for medical products. Galli sees great potential in genetically modified beef, which should prove less allergenic than the regular kind.

In Europe, however, cloned or genetically modified animals can only be used for biomedical research. Its use as food is forbidden—unlike in the United States, where December 2020 saw the FDA authorize genetically modified pigs for human consumption. This is especially relevant for people allergic to the alpha-gal carbohydrate contained by the meat of several kinds of animals, which means it can find its way into various other foods, cosmetics, and medicine.

That being said, even in the United States the meat of cloned pigs is currently not being sold for food.[9]

Avantea strictly adheres to the rule that cloned and genetically modified animals should not end up in the food chain. The company registers its horses as animals not intended for human dietary consumption. The cloned pigs that are not shipped away are also not sent to the slaughterhouse. They are euthanized and incinerated; Galli calmly explained in the hallway that led to the lab.

A strange rattling could be heard from a nearby holding pen. Upon entering it, we saw a huge pig beating its head against the bars of its enclosure. It was a classic example of the stereotypical behavior caused by stress, loneliness, and curtailment of the most basic needs. There was no doubt as to where the pig was going to end up.

Holding our breath—and our ethical misgivings—we listened to Galli list all the things his company does to animals in order to help other animals, especially humans.

We wondered whether Avantea had already been targeted by animal rights activists. "No, they never invaded the company," Galli shook his head. In 2017, however, they did throw Molotov cocktails at a nearby Monsanto facility. After that, Avantea equipped all its buildings with cameras and put iron bars over the lab windows.

"The safety of the northern white rhino embryos has been secured," their creator assured us.

In Galli's opinion, the use of assisted reproduction and stem-cell methods for rescuing a practically extinct species has failed to diminish the general prejudice against biotechnology. "The animal rights organizations are traditionally against the use of biotechnology for saving endangered species. They favor protection of the endangered species' habitats, which is certainly a good thing. But when you push things as far as we've done with the northern white rhinos, other solutions also must be developed."

The assisted reproduction techniques that are tailored to the needs of the northern white rhino could also prove useful for saving other types of rhino, like the Sumatran rhino and black rhino. The same goes for stem-cell technologies. To help other animals, like lions and cheetahs, the protocols would have to be adjusted further.

"However, waiting until the very last minute and then placing all our hopes on an expensive and risky solution—like we did with the northern white rhinos—must not become the norm!" Galli raised his voice. "We need to understand how we got to the current situation, and deal with the problems before it's too late. It is better to prevent than to cure."

When we last spoke, Avantea was putting up a great new complex of labs, pens, and grazing fields next to its somewhat antiquated management building. Galli believed the staff and the animals would be relocated to the new facilities by the end of 2022. Since the expansion consumed most of his time and resources, he hoped to get it done as soon as possible so he could refocus on the science—and saving the northern white rhino.

The fact that he and his wife, Giovanna, share the same professional goals also has its downside. A large part of their outside-of-work conversations, for example, are devoted to their joint enterprise. Since doing real science is a matter of one failed experiment paving the way for the next, the couple's three children have already categorically stated they have no desire to follow in their parents' footsteps.

"All they hear about at home are problems," Galli explained regretfully. "Me, however, well, it wouldn't even occur to me to deplore my choice of career."

"So, what drives you, Cesare?" we wanted to know.

"I want to understand how things work," he replied. "It's of course a hard thing to do. But if you succeed, you can make progress."

"Don't you ever get overcome by the existential fear of meddling with Creation itself?" we tried to provoke a shred of self-doubt.

"Meh," Galli coolly waved his hand. "I'm so busy I hardly have time for philosophical questions."

CHAPTER SIX

The Girls Speak Czech

"*POJD' SEM, NAJNO! POJD' SEM, FATU! POJD' ME!*" CAME ZACHARY MUTAI'S cry, as the caretaker escorted the last two off to graze in the early morning. "Come here, Fatu! Let's go!"

Wait a minute. Hold up. Did we hear him right? Didn't his words have a certain Czech ring?

"Zachary, are you talking to the girls in the Czech language?" we asked as the three of us followed the rhinos' enormous swaying backsides.

"Of course. They are, after all, Czech," Zachary laughed in his quiet and winsomely dignified way.

"They are, after all, Czech," we repeated after him. Then we decided to throw some approximately Czech-sounding commands at the girls. It seemed to work! Slowly, like a pair of double-hulled tankers, Najin and Fatu kept an even keel toward their vast open-air restaurant.

Yes, Najin and Fatu are indeed Czech. Having been born at the Dvůr Králové Zoo, they are able to follow orders—okay, or at least guidelines—in Czech.

In 2009, they were relocated to Kenya along with Sudan and Suni. In 2018, when Sudan passed away, the Czech girls of African origin—or African girls of Czech origin?—became the last two.

Now all we can do is wait.

🦏

For a long time, we saw nothing but fields of wheat, poppy, and potatoes: An interminable monotonous drive along the Czech plains, dotted with sparsely strewn settlements that had too many remnants of communism to be described as "charmingly retro." Suddenly it became interesting when we glimpsed an expansive fertile basin in the distance. Shimmering in the heat, the rural eastern part of the Czech Republic transformed from its typical transitional doldrums into the sort of landscape we last saw in central Kenya when we were driving to the Ol Pejeta Conservancy.

Dvůr Králové was Najin and Fatu's birthplace. It was also their place of residence until they were relocated to Kenya. Having spent the first years of their lives there, the Czech Republic was in fact the girls' real home.

At a glance, the Dvůr Králové Zoo looked like a piece of Africa in Eastern Europe—albeit a rather artificial piece, due to all of the kitschy affectations intended to lure tourists to the region's main attraction.

The animals, however, were very real. Even if, upon strolling the grounds, it sometimes seemed that they, too—while surrounded by neon lights, fast-food restaurants, hotels, parking lots, and giant billboards—formed a part of this reasonably-priced reality show.

Our prevalent first impression was certainly that of the bizarre.

In the scorching Czech summer, a visibly bored cheetah played with its long bushy tail. Like a heat-propelled *perpetuum mobile*, a fossa neurotically circled in its pen. The elephants, looking quite resigned to their prisoner fate, indulged in a long bath. An okapi coyly observed the visitors, who crowded around before its green-tinted domicile, a rather poor approximation of the okapi's real habitat in the northeast of the Congo.

Bands of loudly jubilant children kicked around the ball in front of the cages. Young parents, decked out in the most vivid colors imaginable, tirelessly snapped selfies, looking like wingless birds of paradise. A group of pensioners raptly clung to their guide's every word, even though his every other word was "Africa." *Reggae* music blared from the speakers in front of the "African-village-themed" settlement.

For good measure, the waitress at the improvised African kitchen was dark-skinned. The kitchen smelled of cassava. After all, the African experience must be as authentic as possible, does it not?

A large congregation of pelicans, storks, and other water birds fought for space in their tall but much too tight enclosure, whose edges gave off a distinct reek of bird guano. The birds seemed so robbed of their essence it was hard to watch them. In any case, it was clear they were but an afterthought—just another trinket to sweeten the deal. The zoo's priorities, along with its identity, lay somewhere else.

During the summer season, Safari Park Dvůr Králové is visited daily by 8,000 people according to its PR department.

At "Dvůr," as the locals call it, African ungulates and large mammals were always the prevalent wards. Many of them have become endangered in their natural environments. Some of them, like the bongo antelope and the pygmy hippopotamus, critically so.

The zoo was designed with an eye to making the line that separates the pens from the visitors' paths as unnoticeable as possible. Among many other species, the park is home to Reticulated and Rothschild giraffes, Barbary lions, hippos, zebras (some of them of the rare maneless type), gorillas, African wild dogs, buffalos, gnus, impalas, ostriches—and of course black and southern white rhinos.

Until 2009, five northern white rhinos were also in residence. When Fatu, Najin, Sudan, and Suni were flown to Kenya, only Nabire was left behind in the Czech Republic. She died in 2015.

The rhinos are still the zoo's main attraction, along with the giraffes. Safari Park Dvůr Králové—or "King's Court"—boasts the most numerous rhino population of any of the world's zoos. It is also regarded as one of the globe's most successful breeding facilities for rhinos, with sixty having been born so far.

The Czech Zoo remains the only place where northern white rhinos have ever successfully mated and reproduced outside their natural habitats.

Today, nineteen rhinos are settled in this African chunk of the Czech Republic. Six southern whites—classified as near threatened on the

IUCN Red List, and thirteen critically endangered black rhinos. All of them were brought here from various other European zoos in the framework of an exchange program aimed at facilitating successful mating of the animals.

One part of the zoo experience is an open safari, where a special bus takes tourists right among the zebras, the giraffes, and the impalas. The safari featuring Barbary lions—extinct in nature—can be enjoyed in the comfort of the visitors' own cars.

Apparently, "game drive" can also be experienced in the Czech Republic.

It was an unusual feeling, driving from Austria in the general direction of Poland to see some rhinos. But our trip was motivated by more than our desire to visit Najin and Fatu's birthplace.

Another important motivator was the meeting we had arranged with Jan Stejskal, the zoo's Director of Communication and International Projects and the BioRescue project coordinator.

Since the Czech Zoo was technically still Najin and Fatu's owner, Stejskal was the man who had absolute discretion when it came to authorizing how close to the last two we were allowed to get—and when, if at all.

"Of course, each year we do get a few idiots who step out of the car during the safari to snap a selfie with the lions," Jan Stejskal remarked while we were driving around the zoo.

The sporty looking yet chronically exhausted BioRescue coordinator took to conservation after he left journalism. During his twelve years of mostly covering nature conservation issues, he was guided by the conviction that reading good informative articles would help people make rational decisions. He was young and idealistic; he believed in the power of the written word.

Yet after a few years in journalism, Jan realized the power of the written word was rather limited. He determined that despite all his best efforts, it was getting increasingly hard to penetrate the ironclad prejudice of people's opinions. Good, honest, accurate, and ethical reporting was usually not enough for the key issues of our time to even catch the public attention.

In short, he came to realize that the public itself had changed.

"I was always on the side of nature and sustainable development," Stejskal shared in front of the pens where the rhinos spend their winters. "However, during the last year of my journalistic career, I started losing faith in my chosen vocation. I was beginning to lose hope I'd be able to change anything at all. Especially regarding the environment and especially here in the Czech Republic."

Exiting our vehicle, we headed for a rickety picnic table behind the pens. "I think my breaking point was the Earth Summit (United Nations Conference on Sustainable Development) in Rio de Janeiro in June 2012," Stejskal continued as we sat down. "It was such a huge disappointment. Just a bunch of politicians yakking away, and it was so obvious that not one of them cared! It was sheer pretense and nothing got done. So, I began searching for new alternatives. Words are simply not enough."

Naturally enough, Stejskal's personal renaissance unfolded in stages. An especially important one was when, fed up with urban chaos, the dedicated hiker and runner decided to relocate his family from Prague to the mountain village of Harrachov, located less than an hour's drive from the Dvůr Králové Zoo.

By lucky coincidence, we caught up with Stejskal right after one of his frequent returns from Ol Pejeta, where he had taken part in a meeting of the BioRescue project's key players.

Over the past few years, Kenya had become his second home. It is where his family takes vacations, even though Stejskal had just returned from one of his Kenyan business trips. His addiction to the country was obvious and not at all hard to understand.

"I love the wild," he smiled. "I adore open spaces. Peace and quiet and freedom. I can only hope we will still have some wilderness left in the future—the kind of wilderness not directly impacted by humans. But I fear climate change will have some truly horrendous consequences."

For a while, the three of us sat together silently observing the animals. Their safety was only guaranteed by strong iron fences: existence before essence, surviving before living. Sort of like in *The Matrix*, it occurred to us.

But Jan wasn't fully with us. Confronted with the dictates of his insanely demanding job, he kept scrolling through his phone.

Many of the animals in "King's Court" are descended from ones brought over from Africa by Josef Vágner, the zoo's managing director and spiritual father between 1965 and 1983.

Founded in 1946, the zoo was built close to the rather depressing socialist satellite town of Dvůr Králové by the Laba river. Before Vágner seized control, the zoo was mostly used for showcasing the local fauna, with an occasional lion and polar bear to boot. Educated in forestry, Vágner was passionately in love with Africa. His greatest achievement was bringing over large groups of wild animals, mostly from Czechoslovakia's fraternal African socialist states.[1]

In the photographs displayed in his memorial room at the Dvůr Králové Zoo, Vágner is somewhat reminiscent of Sean Connery. Over his seven expeditions to various African countries, he managed to catch more than 3,000 animals. Some 2,000 of them were imported to Dvůr Králové. The rest were sent to other European zoos.

His hunting companions included his Czech colleagues, hired hunters, and even employees of the *Chipperfield* circus from England. Yet, judging by his diaries, Vágner had caught all 3,000 animals himself. He was certainly the king of "King's Court," nurturing his own cult of personality like a medieval strongman. His memorabilia is still present all around the zoo. Much like his ideas, a few of which became fully implemented only after his death.

Even today, Vágner's relocation of African animals behind the Iron Curtain is considered as the largest relocation of African animals in history. The legendary director's exploits were all the more exceptional as they took place during the most brutal period of communism, when every trip out of Czechoslovakia had to be authorized by the authorities.

Even though Josef Vágner was a member of the communist party, Stejskal claims he was never "a communist by heart." He was, after all, publicly opposed to the Soviet invasion of Czechoslovakia. In those days, even the most renowned Czechs got "cancelled" if they didn't immediately bow down to the pressure of the red Big Brother. Emil Zatopek, one of the greatest athletes of all time, was just one example. Yet the resourceful Vágner managed to hold out for an unusually long time.

Relocating wild animals—including some of the largest ones—posed a tremendous logistical challenge. One of the photos in the Vágner memorial room shows a ship covered with giraffes in wooden pens. At present, transporting animals from Africa to Europe takes about thirty hours. Back then the trip took five to six weeks.

The giraffes and rhinos caught in, say, Uganda were first transported by train to Mombasa. Once there, they were loaded onto ships bound for Hamburg, where other ships took them along the Elba River until they reached the Laba River.

Vágner liked to boast that he only lost around 3 percent of the animals during transport. This was quite an achievement, given that, on average, 30 percent of all transported animals were lost at the time.

With all of the publicity surrounding the new arrivals, the zoo was quickly turned into a major tourist attraction. During the 1970s and the 1980s, up to 20,000 people would visit the place in a day, more than the entire population of the nearby town. Vágner also made enormous gains in influence and public stature.

Even during his expeditions, however, he was criticized for depleting the African natural wealth and for locking wild animals up in zoos, which some still considered to be unethical at the time.

Vágner felt otherwise. He knew African nature was damaged. He had seen it with his own eyes. He also realized that some species could only be saved through special programs of captivity breeding, with the offspring being released into nature when things got better—*if* they got better.

Vágner's brand of nature conservation was one Jan Stejskal could relate to.

Jan Stejskal first visited the Ol Pejeta Conservancy in 2012 as a journalist. He came to check up on the four northern white rhinos that were relocated there from the Czech Republic. The visit represented a key point in his professional pivot.

After returning from Kenya, Jan wrote an article which was very well received in his home country. He raised the question of why, even three years after relocation, the northern white rhinos were refusing to mate—which, after all, had been the relocation's one and only goal.

Stejskal was promptly called up by the zoo's new managing director Přemysl Rabas, who wanted to make some comments about the article. The two of them had such a long and constructive discussion that they decided to stay in touch. Stejskal was of the opinion that the zoo urgently needed some changes, especially in the public-relations department. He listed several recommendations to the director and was promptly repaid with a job offer.

He left journalism for good in 2014, after being completely sucked in by his new responsibilities at the zoo. He became involved with the project of saving the northern white rhino. One of his first moves as the project's coordinator was Europe's first public burning of rhino horns. The occasion received a lot of media coverage—which, in Stejskal's opinion, is key for raising awareness.

As a freshly employed project coordinator, he first travelled to the San Diego Zoo and research center, which boasted its own northern white rhino population at the time. In his own words, Stejskal wanted to set up "a bridge of cooperation," which he deemed crucial for the project's completion.

Yet the mission was hardly a success. The California zoo was not an institution with a great deal of interest in collaboration. Our own experience was rather similar. When we applied to interview the leading San Diego scientists for the purposes of this book, they wanted to charge

$350 for every hour of conversation. Nothing remotely comparable ever happened to us in our entire journalistic careers.

Our communications with California were swiftly brought to a close.

"For a long time, I couldn't wrap my head around the fact that the largest group of northern white rhinos could be found living in a small Czech town," Stejskal went on as we resumed our tour. "The explanation, it turned out, was quite simple and could be spelled out in two words: Josef Vágner."

Saving the northern white rhino, of course, didn't begin with the Bio-Rescue project. It began in 1975, when the last rhino group was brought to Dvůr Králové from Sudan.

"They tried to get them to mate," Stejskal explained. "The eighties saw the beginning of research focused on their reproduction. The zoo employees were studying rhino ovulation cycles. They regularly took urine and sperm samples. Following Fatu's birth in 2000, our zoo started collaborating with the Leibniz-IZW institute in the field of assisted reproduction. Artificial fertilization was tried on Najin and Fatu as far back as 2006. It didn't work. Then, as you know, in 2009, four of our northern white rhinos were relocated to Africa."

It was only around the end of the millennium that many of the world's zoos started switching over to a conservation focus. Dvůr Králové, however, had already been fighting the good fight for more than three decades.

Thus far, the zoo's breeding programs have produced sixty rhinos, 280 giraffes, more than 500 antelopes of the critically endangered *Kobus megaceros* species and more than 8,000 ungulates in total. The zoo is also involved in the conservation breeding program of the critically endangered black rhino.

One of the most intimately familiar with the reasons for these triumphs is Vágner's daughter Lenka Vagnerova, one of his five children.

We visited her at her home on the outskirts of the small town clinging to the safari park, where she runs a bed and breakfast.

The house's interior looked like a museum of her famous father's life. "He was certainly a visionary," she nodded proudly in her meticulously kept garden. "And it was precisely because he dared to step out of the gray average mindset that the system took its revenge on him."

In the words of his daughter, the incident Josef Vágner never fully recovered from took place in 1975.

One night when the legendary managing director was away on an Indian expedition, state veterinarians marched into the Czech Zoo and ordered the deaths of forty-nine giraffes. They claimed the giraffes were suffering from foot-and-mouth disease, which could potentially spread to other animals. This was never confirmed.

In 1983, Vágner finally took an early retirement, exhausted from all of the pressure heaped on him by high state officials. He was also suffering from the consequences of many illnesses and injuries sustained during his African hunting years.

"Above all, he was fed up with having to constantly battle the authorities," Lenka related. During our conversation, she was endearingly forthright about her adoration of her late father.

Near the end of the 1960s, Josef Vágner learned of the northern white rhino species—highly endangered in nature, with less than a dozen specimens residing in the world's zoos. He decided to bring some of them to Dvůr Králové.

He got in touch with Richard Chipperfield, a large-game hunter who negotiated the hunt of six northern white rhinos with the Sudanese government. This was the 1973/1974 hunting season. In the end, seven northern white rhinos were caught. Six were transported to Europe. Among them was a two-year-old calf Vágner decided to name Sudan.

The little rhino reached maturity and grew old as the last living northern white rhino male; he became both the sad embodiment of rapid animal extinction and a global conservationist icon. He died in 2018 in Kenya.

Sudan's skeleton is exhibited in the cellar of Villa Neumann at the Czech Zoo. The sight of the skeleton in the villa's dark cellar is a shocking one—though very easy to miss, given the exhibit's scant markings. It looks like a prehistoric image, something infinitely remote. Yet stripped of context or commentary, it is but a skeleton, and not what it should be: a physical document, an unforgiving testimonial of genocide.

A testimonial of wars, murder, and greed. A testimonial of what we've wrought.

The decision to move the northern white rhinos from the Czech Republic to Kenya was not unanimous. Some of the key players, the Dvůr Králové Zoo being one of them, were in favor of the rhinos remaining in the Czech Republic, where additional efforts could be put into attempts at artificial fertilization.

"Our colleagues, however, especially those in Kenya, were pushing for the move to Africa," Jan Stejskal recalled, beaming genially at some giraffes crossing our path at the safari park. "They believed that the new environment with its huge open spaces and stable climate would stimulate the rhinos' mating. In the beginning, this actually seemed to be the case. But unfortunately, none of the females got pregnant. My job was to mediate between the differing views and steer our team toward the common goal, which, of course, was the species' survival."

Several generations of the Dvůr Králové Zoo's staff have invested enormous amounts of energy and passion into saving the rhinos. "As the official owners, we have a special responsibility. With all the scientific techniques at our disposal, I simply don't understand all the criticism in the vein of: 'Does it even make sense to try to save the species?' I mean, of course it makes sense! If we quit now, we would be betraying our predecessors. Had Mr. Vágner not brought the northern white rhinos from Sudan in 1975, there would be none left. We must not forget that. And so now it's our turn to try to save them," our former journalistic colleague pointed out.

"It wouldn't be right for a single player to make all the decisions. No! An agreement needs to be reached between all the key participants. One day, we'd definitely like to see the northern white rhinos back at our zoo, but that moment is still far away. The current objective is to set up a northern white rhino population in nature. A lot of work will be needed. And also, luck. In order to set up a stable population, part of the rhinos will have to live in the zoos. Over here, we have the entire infrastructure available, including surrogate mothers. One needs to prepare for several years in advance," continued the chief strategist of the project that could change the future of conservation and could provide one for the long list of critically endangered species.

The rhino caretakers at the Dvůr Králové Zoo were certainly opposed to relocating the four northern white rhinos to Kenya. This was what we were told by a white-haired man named Jaroslav as he fed clumps of freshly mowed grass to a trio of southern white rhino females.

Jaroslav has been employed at the zoo for the past fifteen years. "What does working with these rhinos mean to you?" we asked, and we could understand the reply without Jan having to translate it: "Everything."

Jaroslav's every move radiated perfect loving devotion to his wards. The animals he spent most of his time with formed his second family. It was much the same attitude as that of Zachary Mutai, Najin and Fatu's caretaker at Ol Pejeta.

And like Najin and Fatu, the three horned damsels at the Dvůr Králové Zoo were predominantly interested in food. A single one typically consumes between forty and fifty kilograms of grass or hay per day, with fruit serving as dessert.

"Each of the zoo's rhinos has its own character," Jaroslav said dotingly. The southern white rhino named Jessica, who is about to turn twenty-five, is the most congenial. The five-year-old Gaja, on the other hand, is sometimes moody like a teenager.

"They're both okay," Jaroslav said with a curt nod. "But the third one—well, she often gets kind of nervous."

We were separated from the rhinos by heavy iron bars, yet the way Jaroslav stroked and petted each one after their meal was quite similar to how one would pet very large dogs. Following his example, we started stroking the rhinos along the crevasses between their hind legs and stomachs. In perfect Zen-like bliss, the girls would lift the petted leg upward and slightly to the side, keeping balance on only three legs.

The thirty-two-year-old male Kusini, however, called for a somewhat different approach.

Stationed in a nearby pen and weighing more than two tons, he was brought to the zoo in October 2020 from the German park Serengeti at Hodenhagen. The purpose of Kusini's arrival was to spice up the sexual activities of the local southern white rhino females. To put it lyrically, he had come to the Czech Republic to entice and to arouse.[2]

His sheer physical dimensions separated the enormous animal from the other rhinos at the zoo. He also came with his own "police record."

Kusini's infamy was based on a video clip recorded by a visitor to the German park, who caught the rhino assaulting a caretaker's car, pushing it around like an empty cardboard box. The caretaker got off easy, with only a few minor scratches. The vehicle, however, was sent straight to the junkyard.

The ineffably huge Kusini doesn't exactly go to great lengths to hide his character. He is quick to show how he feels about human beings. He gets riled up by every sudden little noise. However, he often doesn't mind being stroked.

His caretaker handed over a large shoe brush and showed us how to best scrape off the dried mud from the two-and-a-half-ton behemoth. Kusini opted to go along with the pleasure afforded to him by the process, until one of the nearby brooms accidentally smacked down to the ground.

Earthquake!

For several years before Kusini's arrival, the southern white and black rhinos at the Czech Zoo failed to produce a single offspring. The caretakers were of the opinion that their wards either lacked privacy or had grown fed up with each other. In 2020, Kusini and another newly arrived black rhino male were supposed to change that. The zoo's staff were convinced it was only a matter of time until the rhinos started producing young ones again. Even a single new arrival would have been treated like a miracle.

The miracle finally happened in March 2022, when a critically endangered black rhino was born—a priceless rarity even for the zoo with the world's largest population of these animals in captivity. Only around 800 of them can still be found in nature. The Dvůr Králové Zoo currently holds fourteen.

The calf was named Kyiv as a sign support for Ukraine in the wake of Russia's aggression. A little less than two weeks after birth he weighed fifty kilograms, gaining a kilogram or two with each passing day.

Given the reproductive problems described here, it is sort of ironic that so many Asian buyers still view rhino-horn powder as a great investment not only into the strength and duration of their erections, but also into their overall fertility.

It seems a foregone conclusion that the rhinos residing at the European zoos are safe. The reality is that their horns are indeed well protected, yet the possibility of a plundering incursion can never be completely ruled out. One of the feared eventualities is that of a helicopter descending straight into the pens. Exaggeration? In 2011 alone, seventy rhino horns were stolen from European museums.

High risk, high reward.

In Jan Stejskal's view, rescuing the northern white rhino is far from only a scientific or conservation project. A great deal of sensibility is needed to harmonize all the various players—both institutions and individuals—while also taking heed of the finances and politics.

This harmonization is Stejskal's job, a job he occasionally loses himself in. Not only on account of the enormous amount of work, but also Stejskal's accentuated need for control. So much responsibility has been placed on his shoulders, especially regarding media relations. His job therefore carried a steep learning curve when it came to being able to say *no*. Judging by our own experience, we can attest he became extremely good at the task.

After leaving journalism behind, Stejskal was forced to develop a number of new skills. "I had to sort of recalibrate my entire personality to fit the new requirements." He now sees this as an invaluable part of his personal growth, the key value in his character.

An additional advantage of his new line of work is its immediate and measurable effect on the world. He may have learned a great deal, yet the basics of his job remain the same: finding the best information available in order to seek out solutions worth presenting to the public. We could only agree that had undergone a sea change over the last few years. The sheer instantaneousness of our age—along with its demands for immediate gratification and the dictatorship of (a)social networks—has long started gnawing at society's core.

"That's part of the reason the only future I can see for myself is in conservation. If we manage to create a baby northern white rhino, it will be mission accomplished and on to the next project. So many tremendous challenges lie before us," Jan Stejskal shared as we pushed through the zoo's summer crowds.

"I can't pretend I share a close connection with Najin and Fatu. They usually don't even recognize me when I visit, but I still love them very much. I'm always so glad to see them. The ones who share a real connection with them are their caretakers, who spend more time with them than with their families," Stejskal continued. He added that in his profession, it was sometimes best not to get too emotionally involved with the animals, since those sorts of emotions can greatly impact one's decision-making.

Priorities must be kept in order all the time.

"We're trying to save a critically endangered species and create a new population," Stejskal summed up the situation. "At the same time, we have to consider the well-being of every individual animal. It's not that the northern white rhino proved an evolutionary failure. Quite the contrary: it was us, humans, who brought them to the brink of extinction. For the rhinos, we are what the meteorite was for the dinosaurs. So now it is our responsibility to save a functionally extinct species."

At fifteen, Jan Stejskal experienced the velvet revolution and the fall of communism. "The revolution took place in my most sensitive years when one's personality and worldview are formed. I loved every minute of it. All of a sudden, we had a president who hung out with Frank Zappa and the Rolling Stones! During those heady days, I also fell in love for the first time. Everything sort of clicked for me. Most of all, I wanted to travel. Until I turned eighteen, I had to respect the rules of living with my parents. Then I set off on my journey."

Stejskal, who loves art and has also written a book about his native region, is still on that journey. "There is a wildness in us and it is best expressed through art," he shared as he cooed to one of the southern white rhino females, who was dreamily leaning onto her pen's door.

"Jan," we asked him, "do you believe in God? After all, saving the northern whites seems a project so daunting it will be hard to pull off without the assistance of some higher, divine force."

"I don't know what to tell you," Stejskal shook his head after a moment's reflection. "I don't want to say no, and I don't want to say yes. I just don't know. At the top of some mountain peak, during one of my highly reflective phases, I might even answer in the affirmative. I might tell you that I indeed felt connected to a kind of transcendence—though nothing like God in his classical form."

All three of us simply nodded at each other. There was nothing to add and nothing left to ask.

Yet Stejskal quickly shied away from metaphysical subjects and glanced down at his sporty sneakers. He told us that during the closing months of 2019, he had come to the brink of a very serious burnout.

The pandemic, what with the travelling bans and working from home, almost came as a blessing.

Jan was finally able to find some time for himself. For the first time in six years, he was able to get enough sleep and also some long-overdue quality time with his family. He afforded himself the unexpected luxury of resuming his reading curriculum, which used to be set at fifty books per year. He reread Marquez's *Love in the Time of Cholera*. But by the time of our visit, the leisure of those months was but a distant memory. Stejskal was again operating at full capacity.

"So, what is it that you most need in life?" we asked our Czech brother in arms.

"Time," he said simply, with a hint of a forlorn smile. "Time for myself and my family. But even more importantly, time for Najin and Fatu, who are so clearly running out of it."

As we roamed the zoo, our mood kept shifting between glee, inspired by the animals' antics, and sadness, even guilt. Endless questions kept buzzing through our heads.

Is it ethical to sacrifice individual members in order to save the species? Could these creatures, born and bred inside a zoo, still be described as wild animals? What sort of message is being conveyed by the existence of these zoos, which have almost literally become the ivory towers of our age? How is it possible that there are serious thoughts of putting the eventual male offspring of Barbary lions to death—a species already extinct in nature—since zoos claim they already have too many?

Are we merely buying indulgences? Are we rescuing our own victims?

Is it a modicum of our conscience we are really trying to save?

CHAPTER SEVEN

What Can Be Sacrificed to Save a Species?

"NAJIN LOOKS POSITIVELY RENEWED! IT'S AS IF SHE'D BEEN INJECTED with some rejuvenating serum!" Photographer Matjaž marveled at Zachary Mutai when we returned to the Ol Pejeta Conservancy in February 2022.

During our first visit, ten months earlier, Najin seemed very lethargic and subdued. Now instead of dragging herself from the pen to the grazing ground, she positively trotted—within, of course, the bounds of rhino maneuverability. She was certainly a lot more agile and vivacious than in April 2021. For an aging aunt, she carried her enormous backside from one uneaten islet of grass to another with surprising ebullience.

"She's in much better shape since she was retired from the oocyte-collection program," Zachary nodded happily. "She is a joy to watch!"

We mentioned our observations to Stephen Ngulu, Ol Pejeta's head veterinarian. He concurred that the retirement had done Najin a world of good. Yet he was quick to add that this was not the sole reason for her improvement. Her mood had also been greatly boosted by the company of a pair of southern white rhino females and especially the castrated male, Owuan—even if they only get to hang out with an electrified fence in between.

"We mustn't forget the northern white rhinos are highly sociable animals," Ngulu pointed out. "Relationships mean a lot to them. They welcome the introduction of new animals into their lives."

The BioRescue consortium made the decision to remove Najin from the oocyte-collection program due to the ethical risk assessment provided by the team of the Italian moral philosopher Barbara de Mori.

It was certainly a decision which made one life much more pleasant, though we can only hope not at the cost of the project's success.

Dr. Barbara de Mori serves as BioRescue's head of ethical risk assessment. We decided to pay her a visit at the University of Padua, where she trains future veterinarians and heads the Ethics Laboratory for Veterinary Medicine, Conservation, and Animal Welfare.

The news of Najin's retirement hardly came as a surprise. Some of the project's key members began hinting at the possibility a few months before the decision was finally taken in October 2021.

At her advanced age of thirty-two, Najin was certainly entitled to her retirement. A few small benign tumors were found on her cervix and her womb, plus a large cyst on her left ovary. An additional factor was that the scientists failed to create a single northern white rhino embryo from her eggs. All the successful ones were thus contributed by the much younger Fatu.

Since continuing with the oocyte-collection program could pose a great threat to Najin's health, her involvement was eventually terminated.

This is a case where the well-being of one animal won out over the necessity of saving an entire species, which makes the tale rather unique. It is fair to say that for a good long while, ethics and conservation weren't exactly synonymous with each other.

"Our decision was extremely difficult. In normal circumstances, it doesn't take nearly as long to weigh the risks to an individual animal. But in this case, the single animal represented one half of the entire population. At any rate, it was the right thing to do," Barbara de Mori opened the conversation as we sat down at one of Padua's most distinguished cafés.

The BioRescue consortium reached the decision unanimously. "From the very beginning, animal well-being was the cornerstone of the project. We felt obligated to respect Najin's welfare," de Mori explained. Her team forwarded its recommendation after a thorough analysis of all ethical dimensions, and after a number of discussions with the project's partners.[1]

It was a very emotional moment for everyone involved, including Barbara de Mori.

Her connection with the last two northern white rhinos goes beyond her professional involvement. "Fatu and I share the same birthday: June twenty-ninth! She is of course younger," the doctor of ethics with long curly red hair smiled.

It was then that we noticed de Mori's incredible resemblance to the actress Susan Sarandon.

The technologies of assisted reproduction are becoming important tools in the fight against encroaching mass extinction. These technologies push the boundaries of what's possible regarding obtaining the offspring of endangered species.

Sometimes these technologies are the only solution, especially when the populations in question are highly dispersed or represented by only a handful of individual animals. In those cases, assisted reproduction can affect a greater number of offspring and facilitate the mingling of the gene pools without the need for animal relocation. More recent technologies also enable the exchange of genetic material between living and dead members of a certain species, namely with the help of the biologic material stored in cryobanks.

These same new approaches, however, raise a few questions regarding the ethical dimensions of their impact on the well-being of individual animals. Such concerns are especially pertinent when it comes to wild animals. "They are sort of like black boxes," Barbara de Mori pointed out. "With them, the risks are higher."

In theory at least, research projects involving experiments on individual animals are subject to a systematic ethical evaluation. The experts are obliged to take a number of elements into account. What is a procedure's risk to the animals in question? What are the possible benefits? How much hardship and permanent damage is the procedure likely to cause?

The pros are meticulously weighed against the cons. One of the most important steps is checking whether "the 3Rs principle"—Replace, Reduce, Refine—was being applied. If at all possible, the experiments need to be performed without animal involvement. If that proves unfeasible, the number of animals participating in the experiment has to be reduced to a minimum. The same goes for their potential suffering.[2]

This model of risk evaluation currently serves as the gold standard of ethical assessment of scientific procedures involving animals. It is, however, only rarely performed in research projects focused on wild animals, although it is absolutely vital—especially with procedures involving assisted reproduction, Barbara de Mori is convinced. After all, the types of procedures used with wild animals are often more experimental and less routine than the ones involving regular breeding stock.

Our knowledge of wild-animal welfare is also much sketchier than our knowledge of farmed animals or animals used in experiments. To make matters worse, wild animals are less used to these sorts of procedures and are thus subjected to additional risk and distress.

Among the many initiatives focused on saving critically endangered species, the BioRescue project stands out because it aims to push the boundaries of conservation programs as they are currently defined.

Ethical-risk assessment is therefore paramount.

"We are primarily interested in the animals' welfare. We assess each step of the veterinary process. No detail is too small to consider. The animals are treated on an individual basis, both in the lab and out in the field," de Mori summed up her team's approach as the waiter brought her a glass of *prosecco* and a modest assortment of classical Italian snacks.

The ethical-risk-assessment process also includes assessing the safety and general well-being of the people involved in the project.

"That is an exceptionally important part," de Mori made sure to point out. "Of course, the welfare of the animals is the main objective, but we mustn't neglect our veterinarian colleagues, especially when it comes to these huge international projects like saving the northern white rhinos. So, you see, ethical risk assessment can have several different objectives.

The overall goal is to keep our ethical standards as high as possible. We must insist on integrity and dignity for all involved."

In collaboration with her colleagues from the Ethics Laboratory for Veterinary Medicine, de Mori developed a tool enabling the BioRescue experts to perform their own ethical risk assessments of procedures involving assisted reproduction and stem-cell technologies. These carefully laid out ethical guidelines proved crucial in preparing the project's decision-makers for the possibility of Najin's retirement.

The tool is called ETHAS, or Ethical Assessment Tool. It combines ethical risk assessment with an assessment of the ethical viability of assisted-reproduction procedures. It is based on a pair of control lists with numerous questions that the experts have to answer before undertaking a procedure. The final assessment of the procedure's ethical viability is a combination of both previous assessments. There are three possible outcomes: the procedure is either acceptable, acceptable provided the consequences are mitigated, or not acceptable.[3]

The tool was developed and tested in collaboration with BioRescue's experts on animal reproduction and also with experts from Kenya Wildlife Service and the Avantea company. Based on meticulously prepared control lists, it was eventually determined all procedures involving the project's rhinos were ethically acceptable.

An interesting aspect of the situation is that the ethics experts led by Barbara de Mori are an integral part of the BioRescue project, instead of serving as outside judges who lay down the law on right and wrong.

"Quality communication is one of the most important aspects of the entire venture," de Mori smiled proudly. "I personally believe everyone should be involved in the decision-making process. We're working shoulder to shoulder, both in the labs and in the field. We learned so much from the scientists and I believe they learned a thing or two from us, as well. New ways of thinking and reaching consensus are developed as we go along. It is a highly inclusive process."

According to de Mori, the BioRescue project functions much like a family. As with all families, its members have to invest a lot of effort into listening so that satisfactory compromises can be reached. The Ol Pejeta rangers guarding Najin and Fatu day and night are also considered as family members.

"When I see the rangers, I see devotion," de Mori nodded reverently. "I see the power of the local community. It is a truly exhilarating thing."

As the conversation drew to a close, our scientific companion refused to predict the project's eventual outcome. "I prefer to focus on my own part of the job." The same goes for all members of her team; removing as much noise as possible from the working process remains crucial. That, in itself, can be a rather Herculean task.

"We are all—every single one of us!—trying to do the best we possibly can," de Mori assured us during our second meeting. "The mere fact that Cesare Galli managed to create so many embryos at the Cremona lab is fantastic. You do have to realize, though, that success can only be achieved gradually—in ethical risk assessment as in everything else."

As if jealous of his owner's devotion to the northern white rhino project, her dog Elliot kept competing for our attention from below the table. For a few moments, we all devoted ourselves to the gentle pup, who was similar in appearance to a Miniature Pinscher.

Developing ETHAS took more than a year. The tool is constantly modified and updated according to the developments within the project. Overall, it was a pleasantly novel experience for the Paduan team, which keeps picking up entirely new skill sets and tools as the situation progresses. Something similar could be said of pretty much everyone involved with the project.

The tool has now been fully standardized for rhino use. It was, however, designed with a view to being used for the ethical risk assessment of saving other critically endangered mammals.

De Mori realizes very well that ethical self-regulation cannot replace the judgment of a group of ethics experts from outside. Nonetheless, she is certain that the tool has the potential to contribute a great deal to the project's wider acceptability.

The ethical assessment of conservation programs is normally performed by an outside "authority" which assesses the ethical acceptability of the entire program before its initiation. "In contrast, ethical self-regulation offers the option of constant scrutiny of a project's ethical dimensions. This is crucial for being able to respond swiftly to the detection of potential harm to the animals' well-being."

Like many of her collaborators within the consortium, Professor Barbara de Mori would like to see the conservation community grow a lot more connected and reject the spirit of profit-driven competition. It was clear she was referring to the San Diego Zoo, which used to host its own northern white rhino population and is now trying to save the species with the aid of stem-cell technologies.

When questioned about her views on whether the northern white rhinos formed their own species or merely a subspecies of the southern whites, de Mori expressed no doubts. Of course, they were their own species, she assured us. "At a glance, you can tell by just looking at them and how they differ from the southern white rhinos."

And so, the freshly retired Najin can now munch on Ol Pejeta's grass in peace. She does, however, still play a crucial role within the project as an ambassador of her species. Furthermore, since she is still capable of raising the young, she might yet find some even more important roles to play.

She certainly has the capacity to transmit social knowledge, culture, and behavioral patterns onto the next generation. "True, the northern white rhinos are close relatives of the southern white rhinos, but the two species' cultural habits and methods of communication differ significantly, which means both Najin and Fatu will be essential for raising the rhinos we're trying to produce," Barbara de Mori is convinced.

The Ol Pejeta rhinos may look at least slightly autistic while engaged with their main task, which is of course grazing. However, most of the time, they linger close to the pen holding a trio of southern white rhinos, made up of a pair of southern white potential surrogate mothers and a castrated male named Owuan. The trio normally sticks very close to where Najin, Fatu, and Tauwo are grazing.

The southern white rhinos have been placed into the pen next to their northern white relatives so that the caretakers can record their daily interactions. BioRescue's experts are set on determining whether the northern white rhinos speak a different language than the southern white rhinos.

This information should certainly come in handy when it is time to start socializing the next generation of northern whites.

How will granny Najin accept the latest additions to the northern white ranks? What will she tell them of her time spent at the Dvůr Králové Zoo? What of her father, Sudan, the last of the northern white rhino males? What will she be able to relate about Fatu's birth or their relocation to their new home in Kenya? What impressions of the human race might she impart through the language of her species?

These were just some of the questions that occurred to us during each morning we spent in Najin and Fatu's company.

"It increasingly looks like the northern whites and the southern whites speak the same language after all," caretaker Zachary Mutai informed us.

They also find each other attractive. In fact, Najin and Tauwo, the southern white females seem to be pretty enamored with Owuan. As already mentioned, they both spend large amounts of time loitering by the electrified fence and glancing in his direction. Owuan, on the other hand, is usually busy smarming up to his two southern white female companions, who habitually push him away. To retain some pride, he marks the territory with forceful and widespread sprays of urine as he beats an honorable retreat.

The feisty southern white male of enviable dimensions and musculature, who sports a meter-long pointed horn, was already present when we first visited the Conservancy. The two females, both also majestic animals, joined him in the pen later. One of them may have been pregnant.

The two southern white females, mother and daughter, are two of the next northern white rhino's five potential surrogate mothers. Owuan's role is at least somewhat akin to that of a male stripper. By trying to mount his female companions, he demonstrates their readiness for mating, signaling this would be the right time to transfer the embryo.

Like all the others, the decision to introduce fresh company on the other side of the fence was only reached after a thorough ethical risk assessment.

There was a bizarre twist to the story in the summer of 2022: One of the surrogate mothers got pregnant by the *officially* castrated Owuan. The BioRescue team had to perform another castration, but the southern white rhino baby is well on the way.

The retirement of Najin, who is now experiencing her second youth in the best possible company, has sped up the development of the Bio-Rescue project's second pillar: the creation of northern white rhino oocytes via stem-cell technology.

In order for the northern white rhino to survive, it will not be enough to produce a sufficient number of offspring to ensure the natural replenishment of the species. The second, equally important objective is securing an adequate level of genetic diversity.

This is a goal the merry retiree Najin can still contribute to by "donating" skin samples, which are then used for creating oocytes.

According to Barbara de Mori, Fatu is taking the procedures very well. "With each new sampling of her skin, she is doing a little better," she reported. "Her well-being is ensured. The procedures don't seem to put her under a lot of stress. Fatu's case demonstrates that you can do a lot for a species' conservation by responding in a timely fashion. The story of Najin, however, is a cautionary tale about what happens when you wait too long."

During our last visit to Kenya in February 2022, Ol Pejeta's head veterinarian, Stephen Ngulu, also assessed the two northern white rhinos as being in pretty sound shape.

For now.

Time passes very quickly. With every minute, the odds are a little more stacked against the *resurrection* of a functionally extinct species. Thomas Hildebrandt's team only got permission for oocyte collection from Kenya Wildlife Service in 2019, even though the BioRescue consortium determined that assisted reproduction was the only viable option as far back as 2015.

"Yes, we are in a hurry," Barbara de Mori nodded. "We realize very well that we've lost a lot of valuable time."

She took a sip of her *prosecco*. "We are in a hurry because the scientific literature tells us that a female rhino over thirty years of age is no longer capable of producing oocytes of sufficient quality. But then again, we can't afford to be in *too much* of a hurry, because we need to be very careful," she grimaced. "But if we are to rescue a functionally extinct species, we simply have to take risks. No risk—no progress, no solution, no breakthrough! That's why making these sorts of decisions is so delicate. All the time, utmost attention must be paid to weighing what is right, what is necessary and what risks might lead to the best possible outcome. This goes for every single step of the process."

"Were there any ethical misgivings before you agreed to participate in the project?" we inquired as the café around us got increasingly loud and rowdy.

Barbara de Mori didn't flinch. "Yes. I thought about that every day."

Her most persistent dilemmas stemmed from the very large number of different players participating in the project and that, with such ventures, the participating animals are never given a say. Almost without exception, their fate is decided without consulting them at all.

"Over the past twenty years I tried to make the animals' voices a little more heard. With projects such as BioRescue, this is not really an issue. A lot of attention is being paid to the animals' preferences."

Even before she joined the project, de Mori was already collaborating with the Berlin-based institute Leibniz-IZW, where head of the BioRescue project, Thomas Hildebrandt, comes from. In tandem with local authorities in the South African Republic, they were developing guidelines for the treatment of elephants employed in the tourism trade. In her words, the work was very demanding, since "the elephant industry" was and remains a crucible of conflicting interests.[4]

De Mori went on to describe her experience from the south of Africa, where she worked for several years.

"If you cared about the well-being of South-African elephants, you'd be obliged to conclude it would be best to prevent all tourist contact with them," she grimaced. "But then you have to take into account the needs of the local communities. The money people get for working with the elephants enables them to feed their children. But let me ask you this. If some higher authorities were indeed to prohibit all contact, what would happen to the people and what would happen to the elephants?"

From her tone and an eloquently raised eyebrow, it was obvious that de Mori's question was a rhetorical one. We were only too happy to let her continue.

"We should use the money that comes from elephant-related tourism for the animals' benefit. The animal-rights organizations demand a ban on elephant tourism, but you must realize that the elephants in the Southern African Republic have survived the mass slaughter of the nineties. Compared to that, they are now not faring too badly."

It was her experience from the South-African project which inspired de Mori to form her team and start developing new tools and approaches to the introduction of ethical-risk assessment as routine in various conservation programs.

For the past few years, she has been collaborating with numerous zoos in Europe and elsewhere. Her mission—to keep putting ethical risk theory into practice—has made her something of a pioneer in the field. Despite all her efforts, however, the ethical risk assessment of procedures involving animals remains a markedly underfunded and understaffed scientific discipline.

So how did she, as a moral philosopher, enter the field of human ethics towards animals? Were the animals dear to her from a very young age?

"Oh yes, it was always a matter of love for me," the Italian professor smiled at the simplicity of our question. "My parents told me that I was already fighting for animal rights when I was three. It was something I seem to have been born with. But I didn't study philosophy because of my love of animals, but because of my desire to explore the world through thought."

The ethics of human attitudes towards animals began to consume her only after she got her doctorate in ethics.

In de Mori's words, that was the perfect moment. "Why? Because it was only then that I succeeded in harmonizing the two different voices in my head." She placed her understanding of human ethics as the basis of our relations to animals. In doing so, she drew on her expertise in biology, her second great passion.

It was a long and arduous journey. Combining social and natural sciences kept opening whole new chapters of inquiry. While slowly perfecting her doctrine, she began questioning the way humans had always taken animals for granted.

"I want to give voice to the creatures who'd always been denied one in a world made by humans. That is my fight. When ethics was introduced as a mandatory subject at our veterinarian university, one part of the mission was accomplished. And I can tell you, when the students were offered these sorts of thinking tools for recalibrating their attitudes, they sure pricked up their ears!" de Mori related proudly.

At the University of Padua, her courses in veterinary ethics only start in the fourth of the five years of study. That is the point when the students are the most open to ethical questions, and sufficiently equipped to start tackling them in earnest. De Mori is also the author of most of the course's literature on the ethical treatment of domestic, laboratory, and wild animals.

🦏

"Working at the university is a wonderful experience," the philosopher continued her tale. "Both there and when I'm out doing fieldwork, I just keep learning all the time. Every day brings a new ethical dilemma."

De Mori lives by Lake Garda. She couldn't bear living in the city, so she prefers to keep making the hour-and-a-half commute to Padua. She usually brings her dog, Elliot, along. During her lectures, he wanders around the auditorium and relaxes the students like some social pedagogue of yore.

De Mori also related that veterinarians suffer from the highest suicide rates among all of Italy's professions. This shocking statistic can be explained by the enormous ethical quandaries faced by veterinarians, as well as their daily insights into animal suffering caused by human society. The situation in the United States is very similar. In 2021, the suicide rates of the U.S. veterinarians were 2.4 times higher than the average among other professions.

"Ethical dilemmas cause moral stress," de Mori nodded gravely, "which is the main reason for members of my profession killing themselves. Moral stress—when you know what's right but are powerless to make it happen—acts like a disease. Since your voice isn't being heard, you learn to suffer on the inside. You must meet the demands of your clients at the expense of animal rights. Over time, the effects can be devastating."

A large role in de Mori's personal and professional development was influenced by the American philosopher Bernard Elliot Rollin, who began lecturing on veterinary ethics at U.S. universities near the end of the 1970s. De Mori initiated their collaboration in 2006. For the next fifteen years, they kept up a lively correspondence, jointly perfecting the curriculum of veterinary-ethics studies.

Her mentor, source of inspiration, and close friend, who is widely regarded as the father of veterinary ethics, passed away only four days before our meeting in Padua. When she spoke of him, de Mori could barely contain her tears.[5]

"Bernard was a part of my family," she told us. "Each year, I got to visit his family in the United States. But for the last two years, I couldn't make the trip because of Covid and now I'll never get to see him again. . . ."

It was clear Rollin's death had had a devastating effect on her. "It's hard. We were writing a book on veterinary ethics in medicine. Now I'll have to finish it on my own."

It is worth mentioning that Professor Rollin had also helped de Mori set up her ethics lab at the University of Padua, which currently remains the only laboratory of its kind anywhere in Europe.

At this point in our conversation, it finally dawned on us whom her beloved dog companion Elliot was named after.

De Mori is in perfect agreement with her late friend, who posited that humankind would never stop using animals, so it should make every effort to learn how to use them compassionately.

Her approach is thus markedly different from the approach of the animal-rights activists, who stopped inviting her to their events after she published an article arguing that lab animals should be treated with compassion. Their view, of course, was that no potentially harmful experiments should be performed on animals at all.

"I understand how the activists feel. They are right. The problem is that, regardless of these sentiments, lab animals will continue to exist. And my role, as I see it, is to make sure they are treated as well as possible. Ninety percent of all lab animals are mice. For various reasons, many people are opposed to their use in scientific experiments. But many of these same people also buy mice poison, which causes an unspeakably agonizing death. The funny thing is that I know of very few grassroots campaigns to ban the sale of these toxins. At least in the labs the animals are anaesthetized."

De Mori followed this up with a rueful shrug, indicating such contradictions were unavoidable. So, if you are in the veterinary-ethics business, you have no choice but to treat them as starting points.

She quickly added that it can often be a struggle, finding the balance between the anthropocentric and the—as she calls it—biocentric worldviews.

She remains convinced her best chance of helping animals lies in helping the veterinarians and others who work with them. "Various points of views need to be considered. Being a vegetarian or not is a matter of my personal ethics. What I teach my students on how to help animals and those who work with them, however, is a matter of my professional ethics."

The dividing line should remain clinically clear. Barbara de Mori has enough trouble as it is separating her professional life from her private one. She fills the latter with mountain hikes, aikido, and diving in Lake Garda. The reality is that her job takes up most of her time. So, she has little time for hobbies or "other passions."

"Hmm, I guess at fifty I should maybe start thinking things over," she laughed, the quality of her laughter letting us know that her choices had long been made. "You see, this was also part of the reason for my divorce. My ex-husband decided to lay down an ultimatum. It was either him or my job. But that's hardly the way things are supposed to work, right? If you're passionate, you should be able to balance both spheres. So now I'm single."

"Barbara, can you tell us what the word ethics means to you in twenty seconds?" we tried to catch her off balance, but de Mori accepted the challenge with gusto.

"Ethics is a kind of an umbrella shielding our values, opinions, and sentiments," she shot back. "It is an attempt at defining good and evil. It is a form of reflection, a perpetual examination of oneself. It is the constant struggle to master our basic emotions and keep them out of the decision-making process."

Even after having met our challenge, the moral philosopher wasn't finished. "The main thing is to be able to get out of one's comfort zone and put oneself in another's skin," she elucidated. "To achieve that, you must freeze your emotions and feelings. Mind you, to freeze is not the same as to deny."

"So, we're talking about a long slog from reflexes to reflection?" we wanted to know.

"Indeed. With proper training, this can become a state of mind. It plays a pivotal role in making the right decisions, especially when the situation calls for a swift response."

So where is Najin and Fatu's place in all this? What does Barbara de Mori see when she gazes into their eyes some early morning at Ol Pejeta? Is she able to put herself in their skin? Is it possible for her to treat them as true equals? How can we even begin to know who they really are and what ethics means *to them*?

"Hmm," de Mori paused for a moment's reflection. "These are very difficult and very beautiful questions. You know what? When I look into their eyes, I see our collective guilt and responsibility. When I look into their eyes, I feel the urgent need for the success of our project. It is not nature's fault that the northern white rhinos have gone functionally extinct. No, humans are to blame! And so, we now need to reverse the course of history and start repairing our sins."

During their leisurely, morning-long breakfast of fresh grass—and especially while gobbling up carrots, their favorite delicacy—Najin and Fatu look perfectly at peace, both with themselves and with the tragic fate of their species. In such moments, their existence can be contemplated in perfect isolation from the wider picture.

So how can one transcend one's aggressive anthropocentric state of mind? Is such a thing actually possible? With a species such as ours, how much room can there even be for fair play?

"You see, the ecosystems care a great deal whether we'll be able to save the northern white rhinos or not," de Mori cut short our persistent efforts to find a chink in her ethical worldview. "We need to do everything in our power to preserve the existing biodiversity—for our sake, if for no one else's! What I'm sure we'll eventually be able to do for the northern white rhinos, we will then be able to do for other species we'd brought to the brink of extinction. As things stand, Najin and Fatu are the symbol of everything we've done wrong. But with just a little luck and a lot of effort, they can be transformed into a symbol of everything we might still do right!"

A true friendship.

Veterinarian Stephen Ngulu during his regular weekly check-up.

Ol Pejeta Conservancy before a storm.

Rhino Cemetery in Ol Pejeta Conservancy.

No rhino can survive 24 hours in the wild.

Ranger patrol in Ol Pejeta Conservancy.

On the night patrol with the elite rangers in Ol Pejeta Conservancy.

Drought, a consequence of climate change, is the biggest threat to local population in Laikipia, Kenya.

The price of elephant' tusks (ivory) is rising again.

A dead elephant in Samburu National Reserve, a victim of human-wildlife conflict.

CROW Rangers, instructors of local rangers in Samburu National Reserve.

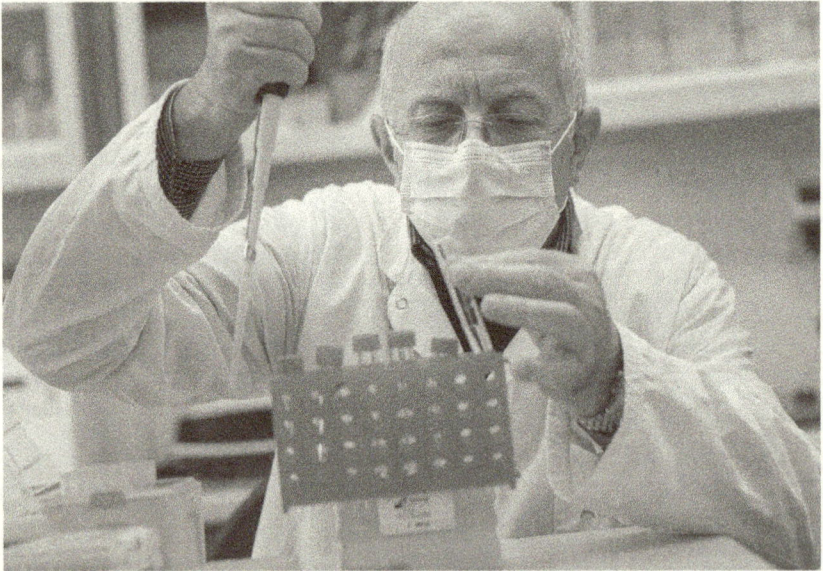

Dr. Cesare Galli in his lab in Cremona with a frozen northern white rhino embryo.

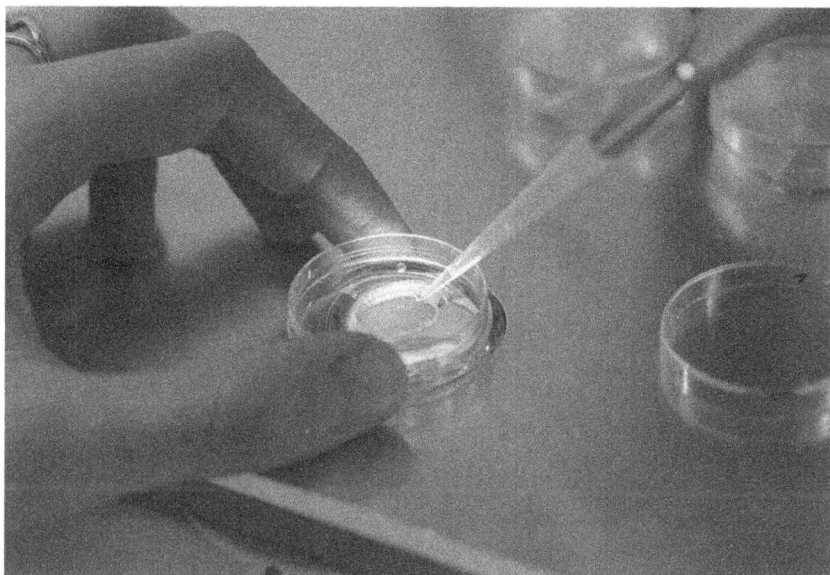

Creating northern white rhino embryos at Avantea company, Italy.

A frozen northern white embryo ready to be transferred to the surrogate mother (a southern white rhino).

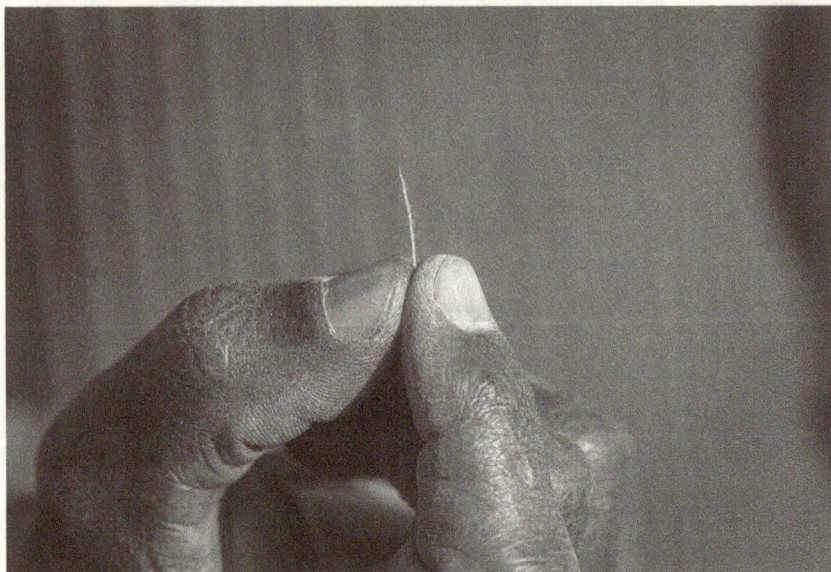

Rhinos' horns are made of keratin, as are hair and nails. It has no medical value.

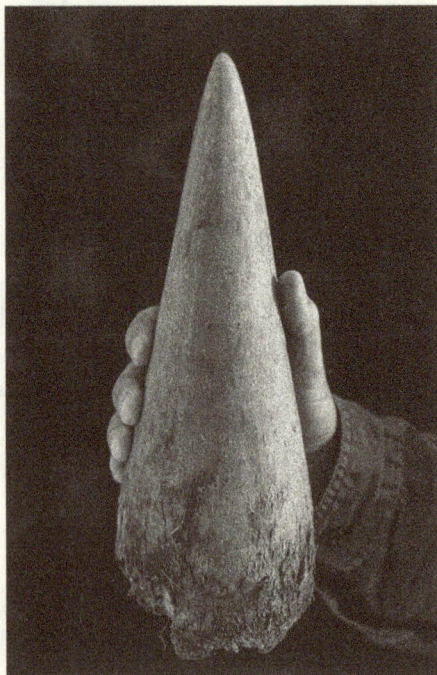

A kilo of a rhino's horn costs up to $100,000 in Vietnam and China.

CHAPTER EIGHT

Resurrecting the Elephants in Samburu

MIKE LESIL, THE RANGER COMMANDER AT THE SAMBURU NATIONAL Reserve, bent over the carcass of a young female elephant. As a final farewell gesture, he gently placed a freshly sprouted twig inside the crater that used to be her skull.

"Rest in peace, dear friend," the forty-year-old ranger whispered with tears in his eyes. Then he averted his gaze from the murdered elephant and looked impotently at the sky.

"Every time one of them dies," he murmured, "a part of us dies as well. A part of me. The elephants are sacred animals. They're our friends, our comrades, our life-companions. We're supposed to be their protectors and yet we failed again. I can't tell you how sad and angry that makes me feel."

Then, struggling with the sheer incomprehensibility of it all, he added: "To us, killing a wild animal is unthinkable."

The young female was murdered by the members of a traditional pastoral community who reside on the outskirts of the Samburu National Reserve, very close to the nearly dried-up Ewaso Ng'iro river.

The unfortunate elephant's crime was to take a stroll across the community's main village. In less dire times, she may have even been forgiven. But not now.

Incensed probably as much by droughts, exploding population, soil erosion, and the wholesale wages of climate change as by the young female, a local shepherd reached for his automatic rifle.

The huge strong animal did not die at once. She managed to cross the shallow muddy river and somehow stumble on to the reserve, where the wildlife is safe—in principle, at least.

Desperately seeking succor, the wounded elephant followed the river downstream, leaving a dark red trail behind. Then she ran out of strength and collapsed. One of the nearby shepherds notified the rangers. By the time of their arrival, she'd already bled to death.

When we were brought to see her remains four days later, hyenas, birds, and worms had already hollowed out much of her insides. Yet the stench of needles and an entirely preventable death still spread far around the sorry mound.

Standing with us by the young female's carcass, Mike Lesil was much too proud to show his tears. We could all but hear him weeping inside.

"I would like to tell the elephant killers that their days are numbered," he spat. "If only I could make people understand, so that they'd stop buying ivory! The tusks belong to the elephants, not to their murderers!"

The hard fact remains that the young female did not die for her tusks. If she died of any one thing, she died of climate change, fueled by much the same appetites that fuel the ivory and rhino horn trades: the unfettered greed, the pathological thirst for dominance, the absence of a shred of self-reflection. All this and more, converges into a vicious cycle and lays waste to entire ecosystems as it gathers steam.

We joined Mike Lesil on his daily rounds along the thoroughly desiccated Samburu Reserve.

"This is the driest part of the year," he explained the situation. "The human-wildlife conflict is at its worst and it gets worse every year. We used to chase Somali poachers, organized crime groups, and local thieves hired by the ivory traders. Now most of the elephants are murdered by the local shepherds fighting the wildlife for pastures and water."[1]

Lesil is certainly one to make historical comparisons, given that he joined the rangers as far back as 2001. The rapid degeneration of the local environment has certainly taken a toll on his worldview. So, having been unable to prevent the torturous death of the young female, Mike Lesil took it extra hard. He was not the only one. The entire ranger unit saw it as a personal blow.

"That was such a rotten day," said a strapping young ranger named Joshua Kakai Lesorogol, while he redoubled his grip on his German automatic rifle. "I'm so mad. What I really want is revenge. Murderers should always pay, right? I want to shoot the guy—just like he shot her!"

As the commanding officer, Lesil immediately made sure to calm the hot-headed young ranger, who had joined the Samburu squad because of his love of nature and especially its wildlife.

Meanwhile, an entire extended family of elephants had gathered in the shade afforded by the nearby trees. The herd was organized around as many as three senior matriarchs. The adult elephants were standing in a circle, providing shade for the young ones resting in the center. The myriad of tracks recently left by the herd testified that the members had recently stopped by the dead female, all of them at the same time.

The rangers ventured that she likely hadn't been a part of their family. Yet, they went to pay their respects all the same.

"Did you know that elephants grieve?" Mike Lesil inquired quietly. "Deaths such as this one hurt them on a very profound level. Sometimes the pain becomes etched into their memories. I've seen too much to believe it could be mere instinct for them. The elephants can feel and think. They're very self-aware, and also highly attuned to their environment."

On average, Mike and his fellow rangers come across a murdered elephant twice per month. It has been a while since he last saw a rhino carcass. That's not because the poachers have grown a conscience, but because they killed the last one at the Samburu National Reserve a few years back.

"Where do the elephants go to die?" we asked the still visibly shaken ranger commander. It was a question long posed by the legendary Polish reporter Ryszard Kapuściński—one for which he'd never found a satisfactory answer.

At first it seemed Mike didn't really hear us, then he snapped into sudden focus.

"Over here, with us," he said. "In the reserve. Here, they feel safe. If they are wounded outside our bounds, then they almost always drag themselves here, providing they are able to. It was what she did—the one the other paid their respects to today."

In light of the young female's tragic demise, it may be surprising to learn that the Samburu National Reserve is really quite a success story, especially given the circumstances: The entire region is heating up and every balance has been thrown out of whack.

The reserve, which is 165 square kilometers in total, is teeming with life. During our stay, Kenya was undergoing a Corona-induced lockdown. While the global freeze on tourism was in place, the local wild animals were afforded incomparably more peace than they were used to.

And so, no matter what direction you turned in, you could see an elephant family roaming the reserve, or a wounded cheetah lying under a tree, recovering from scrapping with a lioness over an antelope. Out of the blue, the late afternoon sun could cast an almost-miraculous sheen over a pair of Grevy's zebras meekly trudging down the strip.

While we observed a pack of giraffes, a large female lion tramped over to our vehicle. "On the hunt for wild boar," Mike whispered in our ear.

All across the parched grassland that was slowly turning into a desert, impalas, antelopes, and kudus foraged for sustenance. They were presided over by birds—such a multitude of birds of every imaginable stripe, shape, and tune! We saw the tracks of a lonesome old buffalo in the mud by the river. "Only six of 'em left in the entire reserve!" we were told by the rangers.

We also got to see crocodiles, we could hear jackals howl, dik-diks and helmeted guineafowl kept jumping out of the bushes as vervet monkey males strutted all over the place provocatively flaunting their light-blue testicles. The baboons, on the other hand, seemed much more interested in browsing through garbage heaps.

When, while indulging in a jog under an incomparable African sunset, you chance across a male baboon or some fresh-looking leopard tracks, your pulse is automatically tripled.

Just the rhinos were missing. There was not a single black or white one in sight. The last member of Samburu's population was killed in 2017.

Yet despite all of the obstacles, things in Kenya are taking a very positive turn—and fast.

Throughout 2020, not a single rhino was murdered in Kenya, which is certainly a historic achievement. The draconic measures passed in 2013 seem to be working. It's little wonder, given that the new dispensation's punishment for poachers caught with ivory or rhino tusks is lifetime imprisonment.[2]

At the same time, several parks and reserves started setting up powerful protective ranger units, who are authorized to use firearms when in direct confrontation with poachers. These rangers are often better trained and better armed than members of the Kenyan army and police. The safari is a key part of the Kenyan tourist trade. Local communities have grown increasingly wise to the fact that the survival of the local animals is directly tied to their own livelihood.

"There may be fewer of them, but the poachers are getting smarter every year," Mike Lesil frowned on the drive to the rangers' training ground on the day after our grisly visit by the river.

"For a while now," he continued, "they've been using the most revolting tactics imaginable. What they do, they seek out a matriarch in the wilds and kill her in as loud a fashion as possible. Her death comes as a terrible disorienting blow for the entire herd. The survivors are at a loss as to what to do, how to behave, where to go."

Mike clenched the steering wheel with barely contained fury. "And so, the terrified and confused elephants gather round the murdered matriarch. At first, they try to wake her up. But they're given little time to grieve, as they are then simply massacred—easy, if you have them bunched up like that. I don't know that there's worse evil than that in the whole world."

Gritting his teeth, the seasoned ranger eased the vehicle into its allotted parking space. The first thing he reached for after he killed the engine was his automatic rifle.

"Well, we'll be keeping watch," he smiled thinly, or maybe it was just a grimace. "For as long as our animal friends keep getting sacrificed to human greed and stupidity, we'll be at war. This is our lifelong mission, saving the lives of our friends."

Mike Lesil's enormous respect for the animals is evident all the time. If one day he were to be trampled by a rogue elephant, he would see it as the most beautiful death imaginable. He would have nothing but understanding for the animal, he shared with a bitter laugh. He would simply give in, his only hope being that it was one of the animals he'd helped protect all his life.

"People have done the elephants such harm," he said, "it's a mystery they're still willing to trust us. We certainly don't deserve their trust. By now, I think we no longer even deserve their cries for help."

One of the reasons behind the distinct drop in Kenya's poaching rates is increasing cooperation between various reserves. With the help of public funding and various private initiatives, conservationists are setting up open wildlife corridors, managing the populations, keeping an eye on endangered species and bolstering local ranger squads.

Given increasingly unforgiving legal measures, the wilds are starting to flourish again. For the most part, Kenya is turning into a shining example of how much can be done.

"Well, we did seem to have come up with a solution," said Abraham Njenga, the Wildlife Department Administrator at the Ol Pejeta Conservancy, and the author of the *I am Justice for Wildlife* blog. "The problem is that a number of countries—including South Africa, Botswana, Namibia, and several others—haven't followed suit. For certain governments and private safari parks, killing animals remains a vital part of the business plan. As long as that is the case, there will be great trouble everywhere, regardless of the local successes."

The Ol Pejeta Conservancy is certainly one success story. By now, it has been turned into a sort of thoroughly fenced-off and highly protected luxury refugee camp for animals rescued from poachers or nearby wars.

Over the next few years, Ol Pejeta plans to link up with at least five other reserves to form a safe-haven network.

"My fear is that the mounting conflict between the human world and the wilds will make this the only workable model of wildlife conservation. The wilds will literally need to be defended," Abraham Njenga shared as we observed a mature southern white rhino wallow blissfully in the mud beneath the massive Mount Kenya.

Much like his comrades at the Samburu National Reserve, Abraham believed that he was on a special mission. Yet he also felt increasingly powerless in combating the quickening force of climate change that had been set in motion in a far-off part of the world.

Abraham would certainly agree with the words of the fifty-nine-year-old Gabriel Leparivo, Samburu's most senior ranger, who had seen and survived everything.

"My hope," Gabriel told us, "is that the people of the future won't only get to see wildlife on television. Humanity is losing its bond with the natural world. Over here, we're still connected to it on a very profound level. That's what makes it so important for people to keep coming to visit. They can see that these animals are real, that they're not just another fiction. They may not get many more chances for that. I am an old man. A lot has changed around here. We are in the process of turning from a savannah to a desert. Rivers have dried up. Water evaporates as soon as it appears. Both the animals and the people are suffering. We've survived many wars, terrorist attacks, and crises of all sorts. But climate change is much worse than the worst of any of that. It makes you feel completely powerless; and even though we're not responsible for it, we seem to be destined to pay the highest price, along with the animal world."

After Gabriel Leparivo took his tea and a small snack under the shade of an acacia tree, he initiated a long monologue about his war with Somali poachers. At Samburu and many other places in Kenya, the Somalis are the first people the locals think of when looking for the culprit of a wide array of mischief. This has been the case for the past thirty years.

"While Somalis undoubtedly were involved in stock theft and took part in poaching and ivory trafficking, they became convenient whipping boys for the government—a community that could be blamed for anything and everything in order to mask official unwillingness to get to grips with the role of political corruption, police graft and criminal networks with official protection in poaching and smuggling," British journalist and writer Keith Somerville writes in his book, *Ivory: Power and Poaching in Africa.*[3]

"In Kenya, things are slowly improving for the elephants although they still remain under threat from humans. Over the last five years, the scope of poaching in East and South Africa was greatly diminished. The drop was so significant that we can no longer say the poachers pose a serious threat to the elephant population in these parts of the continent," we were told by Frank Pope, the CEO of the Kenyan-based research and conservation organization *Save the Elephants.*

Pope, a marine scientist and a former journalist, joined the *Save the Elephants* at the invitation of its legendary founder Iain Douglas-Hamilton.

"So, what has changed concerning poachers and the endangerment of animals?" we asked Pope, given that *Save the Elephants* has operated in Kenya for more than thirty years.

"In 1989, the work of Iain Douglas-Hamilton led to the first international ban on ivory trade," he replied. "Since then, we saw a lot of extreme fluctuations. The seventies and the eighties were simply horrific for the elephants. They were being murdered left and right. By the end of the eighties, the population of African elephants was reduced to less than half, from 1.3 million to around 600,000. In ten years! The international ban on ivory trade was a very positive turn. The elephant populations started to recover."

When that happened, many of Pope's colleagues mistakenly thought the problem had been solved. But by the break of the millennium, elephants were in dire straits again. The problem was that CITES—the Convention on International Trade in Endangered Species of Wild Fauna and Flora—allowed limited ivory trade. For instance, to Japan in 2000, and eight years later to China. This happened to coincide with

China's economic boom,[4] and as a result the demand for ivory exploded, spelling yet another disaster for the African elephants.

"That was a really horrible massacre," Pope was straightforward with us. "We still have to deal with its aftermath, especially in the Central African Republic. But even there, the situation is slowly getting better. We can now reasonably hope the end of the ivory trade is at hand. But I still wouldn't go as far as saying the war has been won. I know very well that's not true. There is still a huge demand for ivory in the Far East and the criminal syndicates are still here. We remain on full alert."

Pope believes that for Kenya to achieve with elephants what has already been achieved with rhinos, a whole new system of ethics will be needed: one based on the concept of co-existence with the animal world. One that is built on respect for the pricelessness of biodiversity.

"Awareness is certainly on the rise," he nodded. "Without global awareness, there can be no sound local solutions, but we also really need a global movement that would fight for the elephants and the rhinos."

Frank Pope places his trust in the younger generations. "We really do need to learn to appreciate animals more. This is my dream, that everyone would simply realize, 'Oh, of course! We need to leave the elephants alone! Their tusks belong only to them!' Well, I know that's not going to happen overnight. We are talking about a paradigm shift, a generational transition. But it is not forbidden to hope that we will see a time when—in Kenya, at least—not a single elephant will be slaughtered by human hands."

Soon after the onset of the Covid-19 pandemic, something akin to panic broke out within the global conservation community. The tourists were gone, and with them a great deal of revenue. With the local armed forces growing less vigilant, things were suddenly looking up for the poachers.

"We were very afraid what might happen to the elephants. We lost a significant amount of funding, but all our programs were still running. The Kenyan areas with the largest elephant populations were hit hardest. A lot of people lost their jobs and that's why hunting for 'bushmeat'—the meat of wild animals—increased," Frank Pope described the impact of the pandemic on the elephants' fate.

"The ivory trade hasn't started flourishing again, thank God. It is now harder to traffic ivory and the surveillance is sharper. But on the other hand, less customs officers are present at the borders. Things function in their own way—mostly chaotically. Close attention needs to be paid to the latest developments," Pope cautioned.

His observations mirrored those of countless other local conservationists: namely, that the horrific consequences of climate change now pose a greater threat to Kenyan elephants than poachers, and that the traditional conflict between humans and wild animals was certain to grow. In fact, half as many elephants died in 2021 in the Samburu/Laikipia ecosystem in northern Kenya from causes recorded as conflict, than died from poaching at the height of the crisis in 2012, we were informed by Frank Pope.

"That's why education and seeking out new models of coexistence are so important," Pope pointed out. "Yes, the elephants are wonderful animals. They should be treasured and protected, but when they ransack the field where you grow the food your family depends on, you suddenly see them in a different light. You get angry, you want revenge. You see, this is about sheer survival!"

More and more people are filling up the planet. Our demographic curve reads like a declaration of war on the remaining animal species—and, lately, on our own kind, as well.

The demand for food is enormous. Arable land is shrinking every day. Water has already been transformed into the oil of the twenty-first century, which is to say the main driver of modern wars.

"These challenges make it imperative for us to find a solution. The problem of overgrazing, which now endangers entire regions, can be avoided. All of the traditional pastoralist tribes graze their cattle precisely where the animal world is at its richest. Something needs to change. The current model is absolutely unsustainable. When we're talking about conservation, we are also talking about conserving ourselves. The situation has grown critical, and we still haven't managed to come up with an appropriate response," Pope kept listing the facts.

"We need to look to the places where the recipe for coexistence has already been discovered," he said. "One hugely important ingredient of such recipes is always that the local communities should not be excluded."

At the Samburu National Reserve, the sun had only just risen above the green hills of Africa, which grow less green with each passing year. The temperature is rapidly rising. The early morning's freshness is becoming just a ghost of a memory.

We decided to board a Cessna plane—the classic old-timer—to soar above one of the most enchanting parts of the world.

For a while, we were unable to utter as much as a syllable. Our British pilot, Angus Carey-Douglas, eventually switched on the elephant tracker used by his organization to monitor herd migrations across vast swathes of Africa.[5]

In Kenya, about 100 elephants are being tracked, seven of them are at the Samburu reserve. Tracking is an important component of the war against poachers. It is also an invaluable tool for the conservation of a species that, in the wake of the continent's dramatic increase in poaching, was recently relegated back into the "endangered" category.

The plane's tracker gave off several increasingly plaintive bleeps. It had picked up the movement of an adult elephant named Edison.

Yet Edison, who liked to travel alone, refused to reveal himself. He was hiding somewhere between the river and the acacia trees. After a few magnificent spins, we fixed our attention on a large elephant caravan. Angus identified the row of ambulant dark spots on the vividly red ground as "The Royal Family."

When we swooped low, the royals understandably closed ranks. The matriarch instinctively protected the young, only to patiently continue the trudge toward the Ewaso Ng'iro River. This year, its serpentine basin was especially slow to fill up.

"This is my favorite place to fly," Angus enthused, handling the plane as silently as possible. "Not only because of the mountains, which always looked like monoliths to me. Or like easels, actually! But it's really down to me knowing the elephants are safest here. Despite all the killing, they are still doing very well. You can see it not only in their numbers, but also in their migration patterns. Here at Samburu, wildlife conservation actually works!"

For the rest of the flight, we were mostly silent. It was simply too beautiful for words. *If God exists*, we jotted down into our notebook, *this is where we came closest to Him.*

As soon as we landed, we had to remind ourselves that such heady conjecture was treacherous at best. After all, elephants were still being shot on a weekly basis.

Lunch time was fast approaching at the Reteti Elephant Sanctuary, an orphanage for elephants located a two-hour-long bumpy drive from Samburu towards the Ethiopian border.

Lunch was served every three hours at the Sanctuary, four times during the day, and four times during the night.

The staff was very busy preparing special "elephant meals": goat milk and powdered milk mixed with hot water and oatmeal. The containers were inscribed with the names of each of the twenty-two elephants that the Sanctuary provides for.

The wards have been divided into two categories, the younger and older orphans. The dividing line has been set at fifteen months of age. The older ones also get to feed while they roam the compound, while the younger ones' sensitive skin prevents them from grazing in the sun.

The elephants are named as soon as they arrive, usually according to the place they are found.

The reasons for them ending up here vary. Their parents might have succumbed to poaching, disease, injury, poisoning, thirst, vengeful attacks by the local communities, veterinary euthanasia, natural causes, or attacks by other animals—to name just a few. In recent times, Kenya has seen a sharp increase of fatal falls into ever more dried-up wells.

The Reteti Elephant Sanctuary is East Africa's first haven for orphaned elephants. It is run as a public institution, managed by the local authorities and kept afloat by a few foreign sponsors.

This is where all the orphans found north of Nairobi are brought. Orphans from the southern half of the country fall under the jurisdiction of the Sheldrick Wildlife Sanctuary, located on the outskirts of Nairobi.

The animals who've found their second home at Reteti can sense when the next meal is coming very easily. With lunch so close at hand, their vast and comfy-looking pen became the source of ever more impatient trumpety sounds.

Translation from *elephantese* was hardly necessary, the message so clearly being: "Please feed us now."

"Every elephant has its own distinct personality," we were told by Dorothy who was in charge of visitor relations at the Reteti Sanctuary.

"In that respect, they're just like humans," Dorothy went on. "They are highly emotional creatures. They can respond both impulsively and in a profoundly thought-out fashion. Again, just like humans. When the young elephants are vexed with something—usually with a human—they get angry and sad and can often take to sulking. Let me tell you, these fellows can carry a grudge, what with their proverbially long memories! They are also highly sociable and playful creatures. Here with us, they quickly get sufficient companionship to help them find inner peace."

Dorothy took us to a room dominated by a schoolboard covered with the names of all of the elephants at the Sanctuary, along with the place they were found and the cause of orphanhood.

"You see, the elephants take care of each other," she continued with motherly approval. "We immediately noticed how our wards show interest in the well-being of each fresh arrival. They're so quick to establish mutual trust. They are exceptionally clever and teachable. The staff here pays a lot of attention to them. We try to give them plenty of the warmth they so clearly need. When they start growing up, we slowly begin to push them away. It's the only way to start preparing them for repatriation into the wild."

As of now, ten elephants have been repatriated in this manner, most of them near the place where they were found.

Dorothy could be forgiven for beaming as she assured us that all ten elephants were now successfully reintegrated. Some of them have found their old families. Some have even been fostered by a new group, something which does not happen very often. Two of the males opted to forge their own way.

"It is a wonderful feeling, knowing you saved a life," she smiled. "All of us here are so glad poaching is on the decline in this part of Kenya. Things are headed in the right direction."

Dorothy has been working at the orphanage for five years now—virtually from the day it opened. She makes no secret of this being her dream job. She not only gets to take care of the elephants, but also the Sanctuary's orphaned antelopes, a lonesome little kudu, and the facility's undisputed star: the charming and communicative giraffe calf named Valentino, who had just turned one month.

Valentino, having not yet been introduced to the concept of fear, kept wandering the orphanage and all but necking with anyone interested.

The Sanctuary's caretakers finally opened the pen gates, just as the hooting and the trumpeting started to smack of open rebellion.

The youngest elephants were the first to be released. The not-so-tiny babies stormed the dining hall at a rapturous full sprint. Some of them immediately locked onto the containers held out by the staff, who kept humming traditional Samburu melodies as the babies fed. Some of the more impatient and enterprising elephants grabbed the containers with their trunks and proceeded unaided. The white mush ran down their faces and feet, but they didn't seem to mind.

Young elephants gain about a kilo each day, two kilos on a good day. Since two kilos is the daily average of what the juveniles gain, the next group at the trough wasn't far behind. The addition of the slightly older group certainly ramped up the chaos factor, but there were no real problems.

Lunch, after all, is the most important meal of the day—even when served eight times in a row.

CHAPTER NINE

The Front Line

IT WAS GROWING DARK FAST. NAJIN AND FATU WERE SLOWLY SHUTTING down for the night. It was so warm for a Kenyan February that, along with their southern white friend Tauwo, they slept out in the open like a band of girl scouts. Before they settled down, they indulged in one last ecstatic mud bath—their favorite activity apart for munching on special treats.

When "the girls" got to sleep, an armed ranger patrol set off on its rounds of the three-square-kilometer pen. With automatic rifles raised, the rangers kept scanning the terrain on the other side of the double electrified fence.

Ol Pejeta has seen firefights with the poachers as recently as 2017. It was then that the management decided to invest in additional security for the 191 resident rhinos. The only way of saving the 191 "walking diamonds" was through the introduction of the so-called green militarization. These types of wars can unfortunately only be won with the help of guns.

"I'm proud to be able to contribute something to the survival of endangered species," a twenty-seven-year-old ranger named David Pirignon informed us, as he stood under a tall watch tower in the gathering dusk.

Pirignon hailed from a village on Ol Pejeta's outskirts. As a trained military dog handler, he joined the ranger squad—75 percent of which had been recruited from the local community—in 2018.

"Not even one of us likes guns. Yet the poachers are organized like the military—or the mafia. We need to be ready all the time," the ardent dog lover told us in the high-pitched voice typical of the Maasai men of Maasai Mara.

"When the full moon is in the sky, we hardly sleep at all," Pirignon pointed up at the third quarter moon above us. "It's when the poachers are most active. So, we are on patrol all the time."

His companion up in the watch tower kept scanning the wider area with the help of binoculars. The black-hooded Pirignon looked like a guerilla warrior as he browsed through the security bulletins on his specially adapted mobile phone. Bursts of static issued out of the rangers' portable radios. The entire unit was clearly in a low crouch, poised for action.

The sky had not yet been illuminated by the stars, but the scene was surprisingly—dangerously—bright, nonetheless.

Another long night lay ahead.

The Ol Pejeta Conservancy is protected by 120 kilometers of dense electrified fence and guarded day and night—from the outside and the inside—by a group of sixty-eight armed rangers. In the northern part of the compound, all of the animals are allowed to cross the Conservancy borders onto the local-community territories—all save the rhinos. Their safety can only be ensured within Conservancy grounds.

This is the rhino's situation all over Africa.

"If the fences were torn down, there wouldn't be a single rhino left in Kenya within a year," predicted Samuel Mutisya, Ol Pejeta's Head of Wildlife Conservation.

"I was never afraid of animals. I feel nothing but respect for them. What I *am* afraid of are bad people," grimaced John Angila during a morning patrol around the Conservancy.

Angila, thirty-six, has been serving with the Ol Pejeta Conservancy Dog Unit since 2017. The Dog—or *K9*—Unit is a special ranger squad tasked with seeking out poachers with the help of trained dogs.

Before joining the unit, Angila spent four years in Ol Pejeta's rhino-monitoring unit. His motivation for joining the rangers couldn't be reduced to the regular and relatively well-paid employment provided by the Conservancy. Angila applied for ranger duty because he wanted to bring some retribution to the poachers, and thereby help protect the local wildlife. Especially the rhinos, who now have a zero chance of survival in the wild.

"You have to be calm, cautious, respectful, and shrewd," John Angila related, striding along the low savannah brushland.

"When we're patrolling the Conservancy—and we do it every single day—we are surrounded by wild animals," he elaborated. "Many lions, hyenas, and buffalos live here. These animals can be very dangerous, especially when they are alone and getting on in years. The elephants and the rhinos, on the other hand, are the easiest to get along with. You can spot them from very far off. If they are bothered by something, they are quick to let you know. In most cases, they give you enough time to retreat in a timely fashion. Though even with them, surprises are still possible. After all, this is the wilds."

A former student of wildlife conservation, Angila quickly realized he would only be able to help effect change through direct fieldwork. "Books make it all sound so logical and solvable," he smiled guardedly. "Well, out in the field, things are rarely so simple."

The seasoned ranger flicked a piece of lint from his military fatigues and checked the safety on his automatic rifle. "It is a wonderful thing to see your work have some real effect and to know that you're actually saving lives; that by taking care of the animals, you're taking care of the entire local community."

The armed guardians of animals in Kenya's national parks and private reserves are authorized to use weapons when their lives are in danger—meaning when they are attacked. Their mandates are reminiscent of Chapter VII of the United Nations Charter regulating the activities of the *Blue Helmets*.

Over the last fifteen years, around 1,500 rangers were killed in Africa. One hundred per year. The exact number of killed poachers is unknown, but it is believed to exceed 10,000. What is certain is that this number includes many innocent people, some of them refugees and migrants who found themselves in the wrong place at the wrong time.[1]

In many parts of Africa, virtually anyone without proper backing can be disposed of under the guise of the fight against poaching.

The last time Angila's unit faced a serious firefight was in 2018. None of the rangers were listed among the casualties. The poachers who tried to shoot their way to Ol Pejeta's black rhinos, however, paid a rather steep price. When the dust settled, four of them lay dead.

Most of the poachers hailed from the vicinity of the Conservancy. One of them was a former employee who had been fired. The incursion was his personal attempt at revenge, we were told by Ol Pejeta's current staff.

The rangers serving in both national and private reserves are required to inform the local police about their activities. There is an official chain of command. The captured poachers can be detained until the arrival of the police, then they have to be handed over.

Over the last few years, a great number of poachers have seen their day in court. Overall, the system is efficient, though it suffers from one significant weakness. The courts have been known to swiftly and draconically dispatch of small-fry poachers, who mostly come from the local environments. Only rarely does the penal system deal with the international-trade middlemen, let alone with the criminal overlords, who are mostly well-connected to the authorities and the dominant corporate structures.

Those few big players who do find their way to court can count on their trials proceeding in a spectacularly slow fashion.

"Intelligence activities are a hugely important part of our work," John Angila related. "We always maintain close ties with the local community. At least three different organized criminal groups are known to operate around our Conservancy. We work in close collaboration with the local police. We train together and since we are so well-armed and well-trained, we often perform duties which would normally fall under policework."

Despite their employment with a private company, the Ol Pejeta rangers are perpetually at the state's disposal as a reserve police force. According to Kenyan law, their armaments belong to the state as well. Several times, the rangers have been deployed for protection services and the prevention of riots.

The rangers knew they would be very busy during the August 2022 presidential elections. Given the traditional friction between various ethnic groups, election season in Kenya is always a highly volatile time. In 2008, several parts of the country were swept in something very similar to a small war, sparked by allegations of electoral fraud.

On the abnormally scorching day we spent in its company, Angila's "K9" unit was training a diligent yet hyperactive dog named Otis and the more sedate Malaika to track down potential poachers.

For this purpose, the Conservancy uses four St. Hubert bloodhounds. Their exceptionally keen sense of smell enables them to carry a wrong-doer's scent for up to seventy-two hours. The unit also includes a female springer spaniel who is trained to find arms and ammunition. There was also a ferocious Dutch shepherd who served as attack specialist. On our second visit to Ol Pejeta, however, we unfortunately learned that he passed away.

All of the dogs are trained for six days each week. The bloodhounds' prize for sniffing out targets is a juicy chunk of meat, the springer spaniel gets a romp with her favorite plastic ball, and the Dutch shepherd got to attack a human being—albeit one wearing a bite-proof vest.

Sunday is the one day of rest for the animals. After being taken for a walk, they are bathed and treated to an extra dose of parasite protection.

To test the efficiency of the professional bloodhound, we were required to hold a rag which John Angila then placed on the black and brown Malaika's enormous snout. The bloodhound was swiftly led away from the scene, while we were invited to find a hiding place about a kilometer and a half away.

Squatting down behind the sturdiest-looking shrub, we didn't have to wait long. The squad led by commander John Tekeles appeared behind our backs in less than twenty minutes. The leashed Malaika took a far-from-hostile sniff and gave a mild bark. She had found us. Her job was done. Tekeles immediately rewarded her with a sausage.

For a few moments, it was dog heaven.

For Malaika, the whole exercise was something very close to a game. Like most dogs, she was highly sensitive to her masters' moods. When the rangers are facing serious action, she turns serious as well—sharp and very sensitive to the surroundings. In those situations, she hardly greets the sniffed-out targets by wagging her tail or licking their faces. Perish the thought!

"Malaika, Otis, and the others are our friends and colleagues. We have the same purpose. So, we work, train and hang out together," John Tekeles went on as he slowly brought our rather routine patrol to a close.

"My motivation for all the hard work is my ambition to ensure quality education for my children," John Angila summed up his life's mission as we were finally able to get some rest in the shade. "Education is key to pretty much everything. It is certainly crucial for developing a proper relationship to the natural world. I want my children to be aware of their oneness with nature. I believe that lack of education is to blame for our destructive attitudes. Most unacceptable of all is how we take nature for granted. Every time I see an elephant, every time I hear a lion roar, I feel both small and infinitely privileged. The animals are worth fighting for!"

"The criminal organizations have eyes and ears everywhere. I don't think there's a single reserve or national park in Kenya where the global crime lords haven't infiltrated some of their men. The same goes for the rest of Africa. And so we have to stay extra careful," said John Djoguna, the commander of Ol Pejeta's special ranger squad.

It was getting near evening, and the sky above the savannah was taking on a metallic blue sheen.

Djoguna's unit was an elite one, charged with additional protection of the Conservancy's rhinos. Organized in two twelve-hour shifts, the unit surveyed the field around the clock. Djoguna sometimes worked for forty-eight hours at a stretch.

He and his colleagues have been trained like elite military forces. Several years ago, the unit bore the brunt of incursions by militias from the neighboring Somalia. The militias were trying to finance arms-purchases by killing animals in Kenyan national parks and selling off the horns and tusks to international smugglers.

The situation is similar in Burkina Faso, which is now very close to becoming a failed state. It was where—on April 26, 2021—the poachers teamed up with the Jamaat Nusrat al-Islam local Al-Qaeda "franchise" to murder the Spanish journalist and our dear friend David Beriain. Also killed were David's cameraman Robert Fraile and an Irish ranger named Rory Young. Young was head of the Chengeta Wildlife Foundation conservation company, which trains African rangers to fight against poachers.

We dedicate this book to David—a veteran of all the wars of the twenty-first century's first two decades. He died in search of the truth.[2]

"The animals see us as a part of nature. We are an integral part of their world and they treat us accordingly," John Djoguna related while he prepared for an evening patrol.

He joined the elite ranger squad five years ago. His initial training lasted only twenty-eight days, but this was because Djoguna already had several years of ranger duty under his belt. Even though Kenya now sees significantly less poaching than during the 2012–2014 period, the time of the pandemic was a busy one for the special squad. As the authorities' attention got redirected to other matters, the doors were again opened wide for the poachers.

"True, we haven't seen a serious firefight for two years now. But hardly a day passes when the alarm doesn't go off, meaning that someone wants to break into the Conservancy. We always respond instantly, so the squad is on duty 24/7," the commander of the elite unit continued as we kept a lookout, laying side by side in the low thicket.

Suddenly, a southern white rhino male started to approach in the dusk. Clearly disturbed by the rangers, the male kept nervously shifting his weight. We were, after all, in his territory.

"We are the ones to always back down. We have no right to cause these animals any undue stress. We're here to protect them, not to bother them," Djoguna explained.

The unit, each of whose members had at least a few fascinating close-encounter tales to tell, swiftly switched location. The rhino followed us for a while, gearing his pace to our own. Twice, he initiated a warning charge. When he was sure we were retreating, he halted and devoted himself to a late supper.

The hour had grown dark and surprisingly cold. Only a few minutes later, our presence attracted the attention of a very powerful elephant male, who started to approach. We must have done something to rile him up.

"You don't run from the elephants. You make a slow and peaceful retreat, unlike with lone buffalos—old, defeated, and cast out by the herd. You see, they are frustrated and irritable enough as it is. When they spot a human being, their mood normally gets much worse. If I see one of those, I climb the nearest tree as fast as I can," John Djoguna laughed.

"I want to work as a ranger until I'm fifty, and then I would like to retire. My one wish is for my children to be able to observe all kinds of animals in their natural environment, instead of in protected enclosures," the ranger commander grew verbose soon after we entered the thick of night, when all we could do was guess at the nature of the various animal noises all around us.

"My other great wish," Djoguna went on, "is for China, Vietnam, and other places where horns and tusks are being bought to finally realize the horror they are causing. Until then, I'll continue to go out every day—as a protector of animals, nature and human beings."

"On average, the poachers try to break in here once every month or so, which means we're doing really well. Let me tell you, things used to be *a lot* worse! We would see as many as six break-in attempts per month and one of the six was usually a success," we were told by Richard Vigne, who served as Ol Pejeta's managing director until the end of 2021.

He, too, was convinced that intelligence activity, based on cooperation with the local community, was key in tackling the local poachers.

"Most of the rhinos now die of natural causes," Vigne reported. "Kenya did a great job in facing the poaching problem, especially regarding rhino and elephant protection from 2014 on. Both populations have been growing by five to six percent each year. As far as the rhinos are concerned, we can safely say the main problem is now lack of space. Meeting their needs requires a lot of room and infrastructure, which is of course expensive and very limiting to the development of the national parks. You see, without sufficient space the species is unable to expand."

The role of Ol Pejeta rangers cannot be reduced to their militant activities. They are also heavily involved in keeping the animals healthy and ensuring their general well-being. They are as knowledgeable about their wards' needs as they are about the local threats to their safety. Besides space and money, this attunement to the animals' needs is the main prerequisite for successful management.

Much of the rangers' attention is directed at monitoring black rhinos, recognizable by the specific pattern of notches on their ears.

"Every single one has to be spotted within three days at the most," Abraham Njenga, Wildlife Department Administrator, explained while driving us around the rain-swept compound. "If one goes missing, we send in reinforcements to comb his or her territory. If we don't find the target within six days, we activate a squad of ten rangers. We occasionally also make use of plane reconnaissance."

The staff is tasked with recording the time and location of the spotting, the rhino's current activity and whether any peers were present.

"On the strength of this information we've been able to define each of our wards' territories," Njenga disclosed. "We know when they're active and when they rest. We know when they like to feed and we know enough about their family ties to draw family trees. We've been gathering this kind of data for ten years now."

The Ol Pejeta staff also monitors the depletion of the black rhinos' favorite snack, the *Acacia drepanolobium* tree commonly known as the whistling thorn, as well as the presence of predators like lions, cheetahs, and hyenas.

The sum of all this information is then used to determine Ol Pejeta's largest sustainable population.

"In 2017 we had a large pride of lions preying on our rhinos. So, we teamed up with the Kenya Wildlife Service and relocated the lions to the Tsavo National Park," Njenga related proudly.

Abraham Njenga got his love of nature from his father, who worked as a safari guide. Abraham began his ranger career at the Tsavo East National Park, where poachers were a much bigger problem than at Ol Pejeta. Luckily, he was never forced to kill a human being.

"Over the last years, the situation in Kenya was much improved," the athletically built and smartly turned-out Njenga related as we observed a female black rhino and her two-month-old calf—Ol Pejeta's 150th rhino resident.

"Some say the rangers shouldn't be armed," the Wildlife Department Administrator continued. "But I can tell you that it was the decisiveness of our responses that made it possible that not a single rhino was murdered in Kenya over the last two years."

Njenga regularly receives job offers in Kenya and from abroad. He is, however, perfectly happy at Ol Pejeta, where he has everything he loves and needs. His favorite times are spent in the thicket—the deeper the better. If his family were allowed to live with him at the Conservancy, the situation would be nothing short of ideal.

"It is so wonderful to be part of a project which could shape the future of conservation. Despite all the dangers, our work can also be extremely fun. There is a hierarchy of courage among the rangers. We've all seen someone who'd soiled his pants before a charging rhino," Njenga related with a merry laugh.

At the Samburu Natural Reserve, located a three-hour drive north from Ol Pejeta, the rangers operate in much harsher conditions. In this drier and less populated part of Kenya, conflicts within the local community, based on traditional grievances and the consequences of climate change, are omnipresent.

Samburu County is an extremely poor one. The worst drought of the century and the fallout from the pandemic have done their worst to exacerbate the situation for the already struggling population. Many of the locals have been pushed to the very brink of survival. Yet the Samburu rangers—under-equipped, permanently burned-out, and completely exposed—are pushing on with their toilsome mission.

One of the job's few perks is constant education. For a few years now, the Samburu rangers have been trained by a varied and colorful array of international instructors.

"If you run this slow, you'll be massacred!" yelled Andy Martin, executive director of the Conservation Rangers Operating Worldwide (CROW) organization. The rangers in training dutifully redoubled their efforts to follow his directions under the scorching sun.

The rangers at Samburu—one of the world's most beautiful wildlife reserves—are manning the frontline in the war against poaching. Their dedication, along with updated legislation punishing the caught poachers with lifetime imprisonment and/or tweny million Kenyan schillings ($150,000), helped turn the reserve into a huge success story.

But the rangers still have to train as if a war could break out any minute, which, in these parts, it certainly could.

"Back in the eighties, when I was starting out as a ranger, we used to chase lone poachers," Gabriel Leparivo recalled. Then the changes in legislature were put in effect, but the entire continent changed as well.

"So many wars have broken out!" Leparivo went on. "Around here, we felt the consequences of the conflict in nearby Somalia. Organized militias of well-armed Somalian poachers began to make raids on our reserve. Those were the most difficult times of all. We were poorly equipped and trained and they had a whole war machine behind them."

We were talking to Gabriel Leparivo, Samburu's eldest ranger, while he observed his comrades train in martial arts. He went on to relate how decades of fighting poachers had cost him several friends.

"Today," he summed up, "the poachers who try to breach our borders are relatively few and far between. We've got that situation under control. We've also really grown as rangers, helped by the training provided by the CROW organization. Now we can not only handle most imaginable situations, but we've also got pretty good at predicting. However, times are still hard. The greatest threat to the local wildlife now comes from within, meaning from our local Samburu community. The people here have begun to see the animals—especially elephants—as direct competition for water and farmland. Elephants are now often killed without anyone even bothering to collect their tusks. Some have actually been murdered out of revenge. Every other week we come across a fresh carcass. It is a horrible, horrible thing."

For a moment, Leparivo stopped polishing the black military boots in his hands and gazed forlornly at the ground. Even this downcast, he looked at least fifteen years younger than his actual age.

After collecting his thoughts, he expressed his conviction that brute force alone could never preserve the precarious balance between humans and animals. Yet he was also delighted to report he had lately begun to sense a great deal of progress in the local community's outlook.

"It's the people, you know," he smiled wistfully. "Only the people here can prevent the parks and the reserves from being slowly turned into open-air natural-history museums. You could say wildlife conservation offers an enlightened business model. With the help of civilized tourism, the locals will be able to survive. This is part of the reason the recent pandemic came as such a horrible blow to us all."

"For years now, I've been visiting at least one primary school or high school each week," Mike Lesil related at the Oryx Airstrip Samburu, a small airport atop a small elevation in the middle of the savannah.

"I tell the children about my work," he elaborated. "I tell them about the importance of wild animals, endangered species, and climate change.

You see, I firmly believe education is the key to peaceful coexistence and, therefore, to our common future."

We were standing in the scant shade provided by an acacia tree and observing the members of Lesil's ranger unit train in martial arts and patrol formations. They were carefully and passionately watched over by three instructors from the CROW organization.

The local rangers were of course glad for the help. They were only too aware that many of the poachers were merciless professional killers: well-trained, well-organized, and well-armed.

The rangers of Samburu have a long history of loss. "We are basically at war," Mike Lesil acknowledged. "And so, we train for war. That's why we're so indebted to these international instructors for all their help. We're taking the poachers on as best we can, but we noticed they'd been training up as well. We are pitted against big bucks, corporations—even states. For a long time, we have been mostly left to our own devices. But things are a bit easier for us now."

The two dozen rangers, sweating profusely under the foreign instructors' vigilant gaze, included four female rangers. Mike Lesil's ranger unit also employed his wife Lea, but she was currently taking some time off to nurse their one-month-old boy named Joe Biden.

"Around here, we grow up with the wild and all its animals," Lesil continued, leading us on toward the reserve's mess hall. "Most of the rangers come from the local Samburu communities. We're intimately familiar with the plight of our people—traditional shepherds who'd spent most of their lives tending their cattle, but so many of our pastures have vanished. They have simply shriveled up. The soil has turned to dust. It's never been hotter and drier than it is now. "

Struggling to survive, the locals started crossing the borders of the Samburu national reserve. The hard fact is that Kenya is one of those especially unfortunate countries where the wages of climate change have coincided with a population boom.

"That's how the conflict with the wild was born," said Lesil, who in his spare time works as a Samburu tourist guide. "The reserve went as far as setting up two special corridors for the cattle, but the conflict only keeps escalating."

With the Corona pandemic wreaking such havoc on the local economy, an increase in poaching was only to be expected, but our new friends at the Samburu were ready. Years of clashing with the Somali mercenary militias have gotten the rangers into excellent fighting shape.

"I really don't like shooting," Lesil shook his head emphatically. "I try very hard to resolve all matters through diplomacy. This is why cooperation with the local community is so important, but there is just no peaceful dealing with the foreign poachers. Again, we are basically at war. They are backed by some very rich and powerful people. But, I can tell you that so far, we've won every single armed battle we've been in. We know these parts like the back of our hands. Also, we're the ones defending our homes and our animals—and, of course, ourselves as well!"

The three foreign instructors who decided to spend their vacations training the local rangers free of charge were not the type to hide before the sun. The veins in their muscular necks kept bulging as they first demonstrated each new drill themselves.

Patiently and attentively, they went through the checklist. Knife-fighting; opponent neutralization; jumping off a speeding vehicle; martial arts; patrol formations; marksmanship; weapon maintenance; offensive and defensive tactics. The list went on and on.

The twenty-two rangers-in-training followed up on the instructors' every word with the utmost dedication. Discipline was high, almost military-like, but the general atmosphere was very relaxed. Above all, there was none of that forced authority and rote obedience stuff. The only things that counted were skill, experience, and inner fervor.

"My profession—my mission—is to protect the weak," said Andy Martin, head of the international instructor team. Originally from Piedmont, Martin spent some time serving in the Italian army and then took to sub-contracting for various private security firms.

The natural world had always been his one true passion. When he learned of the heinous story behind the international ivory and rhino trades, he knew he needed to join the fight straight away.

In 2007 he travelled to the South African Republic to undergo ranger training. He spent the next few years serving in private reserves all over Africa, only to become intimately acquainted with everything that was wrong. The intentional breeding of wild animals for hunting purposes, rampant corruption, and rangers collaborating with poachers were just some of the most chilling examples.

Martin kept learning as he went along, slowly perfecting his ranger's craft. Perhaps most importantly, he came to realize it was pointless to work with conservation organizations led by people in air-conditioned offices and dominated by their marketing departments. Or at least it seemed pointless to him as a sworn terrain-work fundamentalist.

Just as the sun was about to reach its apogee, Andy Martin announced the end of the first half of the day's training regimen. Lunch was served in front of a barely standing edifice clinging to the small airstrip.

Visibly relieved, the local rangers immediately sought shade and started rebuilding their caloric reserves. They also made sure to down copious quantities of water and tea. Hardly ten minutes passed before all of them, every single one, fell asleep.

Since the animals had retired to the shade as well, the only sound to pierce the vicious noonday swelter was an occasional snore.

The team of foreign instructors, however, did not yield to the temptation of a refreshing nap. Andy Martin checked up on the equipment while planning the afternoon second part of the drill. While doing both, he was still focused enough to tell his story.

"I wanted to fill a void," he related. "And I soon got my opportunity. In the United States, where I ran a company at the time, an organization called Conservation Rangers Operating Worldwide sprang into existence. This was in 2016. I knew the founders—hell, I used to work with one of them in Pakistan! In 2017 I was made executive director."

The CROW group is organized around its lengthy roster of volunteers. At the moment, about fifty rangers-instructors are among its ranks. All they ask for when approached by reserves and national parks to help drill the staff are lodgings and help with managing local logistics. The rest is arranged and funded by the instructors, including plane tickets.

"At CROW, we are always looking for highly motivated people—those driven by something other than the desire to make a quick buck. So, I can tell you we've got some pretty special people with us by now. A lot of them come from a military background. Most are obsessed with nature, especially with the wild and all its animals. To be frank, some of them aren't really cut out for civilian life. I can understand that being so much like them," grinned the former MME wrestler and self-educated erudite, who liked to season his discussions of wildlife conservation with quotes from Voltaire and Dostoevsky.

As a private contractor, Andy Martin participated in operations in Sinai and Pakistan. Along the way, he was treated to a crash course in the inner workings of the global economy.

"As for humanity, I have long lost most of my illusions," he shrugged. "I'll be honest with you: the world of men disgusts me to its core. Taking care of wild animals is my way of trying to do something constructive. I really hope these animals outlive us all. My belief is that we're living in the times of the greatest global challenges, well, ever! I'm afraid we may not be up to the task."

Martin, by far the most gregarious of the three CROW instructors present, allowed himself a further personal digression. "Every felled tree, every miserable child and every murdered animal hurt me on a very deep level," he grimaced. "And then all I feel is an uncontainable need to act. This makes me quite the weirdo, doesn't it? I often feel rather lonely, especially when I wonder how come so few see or care about everything that's wrong with the world."

The words kept pouring out of the inhumanly strong Italian. "They say there's no such thing as a just war," he concluded. "But ours is."

With that, he re-joined his two CROW colleagues. The German K.D. was a seasoned sniper and the Italian J.P. was a fellow ex-member of the national military. Laughing in the sun, the three seasoned warriors reclaimed the training grounds for the benefit of the local rangers, whose short naps had done a world of good.

The local rangers live along various entry points to the Samburu reserve, with some situated within the reserve itself. The majority reside on the outskirts of a struggling town called Archer's Post, located by the key road connecting central Kenya to the north.

A decade ago, the World Bank chipped in towards building the rangers' lodgings. By now, however, they are in a rather rough shape.

There is no running water. The rangers' families had to get used to drawing the ever-saltier liquid from a deep well. There is also no electricity, though a pair of small solar panels provide sufficient power to maybe charge a cell phone or two.

This means that the rangers share the life of the local pastoral communities, cut off from the urban centers and often completely reliant on the elements. The consequences of climate change have thus already hugely impacted the rangers' lives. A starving lion occasionally crosses their courtyards, or even an elephant, which means the rangers' duties are increasingly encroaching on their spare time.

"When I applied for the job, I knew very little about poaching and the situation at the reserve," related Eunice Lenyakopiro, one of the four female rangers taking part in the CROW training exercise. "All I wanted was to help protect the animals. I've always felt a strong connection to them."

We were talking to the twenty-five-year-old Eunice, in front of her humble home right before the entrance to the reserve.

"Today, I learned a great deal," she smiled bashfully. "These training exercises make me feel much more confident and prepared to take on those trying to destroy the ones we protect. You know, it's war out there."

Eunice, a former runner, joined the reserve in 2014—just like her husband Joshua Kakai Lesorgol. During our visit, Joshua was still badly shaken by the morning's find of the dead young female elephant, the one killed by the local shepherds with automatic rifles.

"I used to be a shepherd myself," Joshua all but snarled. "I used to come across the wild animals all the time. But I wouldn't dream of killing one! I simply can't understand how anyone would do such a thing! The elephants are our best friends. This is one of the foundations of Samburu culture. Killing an elephant is a taboo."

Joshua Kakai Lesorgol is considered to be one of the reserve's most steadfast rangers. "We need more men," the young man informed us. "Our capabilities are stretched. We're working virtually non-stop. We also need more weapons, vehicles, and gas. We're neither adequately trained nor paid for our work. It's hard. But all that doesn't begin to touch my motivation! I do what I do because of my profound love of nature and also, because I want to provide a good education for my children. Nothing in Kenya is free of charge and it is only in schools that our kids can learn the true importance of nature conservation."

In front of the rangers' dwellings, which are reminiscent of solitary confinement cells at some of the world's most infamous prisons, a brood of emaciated hens picked through the refuse. At the nearby junkyard, a number of skeletal goats kept gnawing at everything at least notionally resembling food. Heaps of discarded plastic dotted the landscape behind the decrepit buildings, which a mere decade ago used to be the pride of the local community. Many of the larger families lived in improvised tents.

The Potemkin-village quality of the entire place was laid bare the instant the foreign investors decided to bail.

There was garbage everywhere. A warm late-afternoon wind kept tossing up the countless dirty plastic bags, while spreading the mixed scent of frugal dinners and burned plastic.

Yet our unannounced visit nonetheless brought a brief jolt of excitement to the small ranger community. In front of their shacks and tents, the girls and women in traditionally colorful tribal garb quickly set up a small marketplace selling clothes and jewelry. Almost miraculously, since there are no shops, the locals managed to come up with bottled water. We were even treated to an impromptu dance show.

"You have to understand how we live. You must see it with your own eyes. Only then will you be able to realize how hard things are for us and how important it is for everyone to protect the environment," Mike Lesil, the ranger commander, said in parting.

"Perhaps . . ." he added. "Well, perhaps we can be assisted by the fact that it is generally much easier to love animals than humans."

CHAPTER TEN

Life After Life

THE NORTHERN WHITE RHINO NAMED NABIRE DIED ON JULY 27, 2015. She was thirty-one years old. Her death was caused by complications following the bursting of the large cyst on her uterus.

Nabire spent her entire life at the Dvůr Králové Zoo. She was descended from Nasima, the most fertile northern white female in captivity, and Sudan, the last of the northern white males. Which made her Najin's half-sister and Fatu's aunt.

When Najin, Fatu, Sudan, and Suni were relocated to Ol Pejeta in 2009, Nabire was left behind in the Czech Republic. She was already suffering from many cysts on her uterus, so it was clear she would never be able to reproduce naturally. All the same, she remained a part of the northern white rhino breeding program. The experts were still hoping to collect at least a few oocytes from her relatively sound left ovary.

When Nabire died, the said ovary was taken out and shipped off to Cremona. The embryologist Cesare Galli managed to collect a few oocytes, as well as to mature and fertilize them. . . . Yet so far he failed to create an embryo from Nabire's eggs.

After her demise, Nabire also contributed samples of skin tissue, the aim being to use them in further oocyte creation.

Eight years ago, this was still pure science fiction. The BioRescue scientists, however, recently made a hugely important step in the right direction. With Nabire's skin cells, they managed to create primordial germ cells—the precursors of eggs and sperm. A world's first for the large mammals.

The final step of the plan is to produce artificial rhino gametes (eggs and sperm) from preserved tissue, so that the scientists can start creating embryos.

Nabire never got the chance to taste the fresh grasses at Ol Pejeta. She also never had the chance to become a mother. But it looks like she will be able to become one in her afterlife.

Biodiversity is a highly abstract-sounding concept, yet it merely means the sum of all living creatures and their habitats. It denotes the diversity of life on our planet—the life we seem so hell-bent on destroying.

The sixth mass extinction is writing a new, decidedly "unnatural" history of animal and plant evolution.

For the conservation of some species, the classic methods of conservation like habitat protection and breeding programs are no longer sufficient. These species' survival can now only be ensured through the application of new technologies. As long, of course, as they are not merely an extension of "the myth of the techno-fix"s—the belief that every problem can be fixed through technology, so there is no need to deal with the underlying causes.

Stem-cell technologies are currently considered as one of the most promising technologies for rescuing endangered species. In combination with advanced methods of assisted reproduction and the use of biologic material in cryobanks, "libraries of life," they raise the possibility of creating new animals from a single skin cell or drop of blood.[1]

These technologies could also greatly contribute to the improvement of overall health and genetic diversity of critically endangered animal populations.

"Only when numbers get so low that the genetic contribution of every last animal (including those represented only in frozen cell lines) contributes measurably to the total species diversity—maybe around 10 individuals—would we want to do everything possible to ensure that those genes are transmitted to future generations. Tragically, northern white rhinos have undergone just such a decimation," Robert Lacy, a conservation scientist at the Chicago Zoological Society and chairman of the Conservation Breeding Specialist Group attached to the International Union for the Conservation of Nature (IUCN), told the BBC in 2011, when ten northern white rhinos were still surviving in captivity.[2]

The northern white rhinos are the world's most endangered mammals. The project of their rescue through stem-cell technologies is therefore the supreme test of these technologies' efficiency.

"It is too late for the northern white rhinos to be saved by conventional conservation efforts, and the only hope for the survival of this species is introduction of new, unconventional experimental strategies. One of these strategies is to take advantage of the virtually unlimited differentiation capacity of pluripotent stem cells using them to generate gametes that could be incorporated into assisted reproduction technologies," the experts of the "American branch" of the project of saving the northern white rhinos, led by the San Diego Zoo Wildlife Alliance, wrote in an article published in the *Stem Cells and Development* magazine in 2021.[3]

As the organism's building blocks, stem cells are responsible for repairing tissue damage and for replacing dead cells. They also possess the capacity for both self-renewal and differentiation into more specialized types of tissue cells.

There are several possible sources of stem cells. Embryonic stem cells are derived from the cells of an embryo aged between three and five days. Embryonic stem cells are pluripotent, which means they are able to develop into every single one of the body's cells, except for placenta cells.

Yet since the process of stem-cell collection is fatal for the embryo, the use of embryonal stem cells has raised several ethical objections. Only a few of the world's countries currently condone this vein of research.

Stem cells can also be collected from the umbilical cord and from the umbilical-cord blood. The stem cells collected in this fashion have long been used for treating various blood disorders. The third possible source are the stem cells obtained from the tissues and organs of adult organisms. Their limitation is that they are no longer able to differentiate into every type of specialized cells, only into certain specific ones.

This last type of stem cells, however, can be modified so that the cells take on the properties of embryonic stem cells. The procedure is called "cell reprogramming" and does not entail the sacrifice of the donor organism.

A landmark study in 2006 demonstrated that skin cells can be reprogrammed into stem cells. The cells created in this fashion are called pluripotent stem cells. Like the embryonic stem cells, they are able to differentiate into any type of cells, including reproductive cells.[4]

Healing on the basis of stem-cell technologies—the so-called regenerative medicine—is a field which has long been promising miracles. Yet over the past decade, only a few such treatments have been approved. Most of them have to do with the treatment of various blood diseases. A number of other stem-cell therapies are currently in the clinical-trial phase.

The research is mostly focused on the potential for treating illnesses and injuries, gathering new insight into disease development, creating replacements for damaged cells. . . . As well as on exploring the possibilities of the use of stem cells in testing the safety and efficiency of new medical products.

The scientists are also probing into other possible areas of application. Tissue engineering utilizes animal (stem) cells in the production of "clean" meat, which requires virtually no suffering on the animals' part. This has the potential to significantly reduce our reliance on animal agriculture breeding. The *BlueNalu* American company already creates "seafood" from the cell lines of marine animals. Another American company called *Eat Just* sells lab-grown chicken medallions obtained from the cells of live chickens.[5,6]

In order to treat infertility, scientists are also trying to unlock the secret of turning stem cells into reproductive cells. The leading authority in this field is Katsuhiko Hayashi at the Osaka University (formerly Kyushu University) University in Japan. In 2016, he was able to create egg cells from the skin cells of mice, artificially inseminate them and insert them into female mice who later bore healthy and reproductively viable offspring.[7]

The scientists of the BioRescue consortium are now collaborating with Hayashi, his colleague Masafumi Hayashi, and their team in the production of egg cells from the skin cells of a far mightier type of animal—namely the northern white rhino.

As far back as 2015, when the plan for saving the northern white rhinos was formed, the BioRescue strategists realized stem-cell technologies would very likely play an important part. Yet until recently, the consortium was much more focused on the rescue project's main pillar, which involves creating embryos through artificial insemination of egg cells obtained from Fatu and Najin.

Understandably enough, Najin's forced reproductive retirement meant the project's second pillar—involving the use of stem cells—started increasingly coming to the fore.

The goal is to create "artificial" northern white rhino eggs from the skin tissue of the species' deceased representatives. Beside Katsuhiko Hayashi, German scientists are also involved in reaching this objective. One of the project's key proponents is Dr. Sebastian Diecke with his team of collaborators at the Max Delbrück Center for Molecular Medicine in Berlin.

"There is of course a lot of pressure. But I am even more motivated by the possibility of effecting positive change. We all know we're working for the higher good, so we're doing everything in our power. If we succeed, we will show that, for the first time in human history, extinction can be reversed. I really think we have a high chance of success," Dr. Diecke told us over Zoom in January 2022.

Sebastian Diecke joined the BioRescue project as soon as it was launched in 2015. The institution he is employed with joined only later—in 2019, when it was awarded €4.2 million by the German ministry of education and research for three years of stem-cell research aimed at saving the northern white rhinos.

Before then, Diecke and his team were for the most part doing the research in their spare time. Out of sheer enthusiasm. The financing was secured through private donations.

"We tried to understand at least a small part of cell communication, especially how illnesses are created and how we can treat them," head of the Pluripotent Stem Cell Platform at the Max Delbrück Center explained.

Diecke's current research is mostly focused on advances in the treatment of various human diseases. His team is delving into the molecular foundations of human health, in order to translate the findings into clinical practice.

"So how different is your approach when you're helping a pair of animal patients on whom the survival of an entire species is hinged on?" we asked him. "Is there a heightened sense of moral responsibility?"

"I am a biologist by education," he replied. "And I also studied veterinary medicine for a while. So, I get very excited by nature's small mysteries. I find the BioRescue project especially appealing because it combines both of my primary fields of interest with my expertise in molecular-biology methods. The northern white rhinos have not gone extinct because of some flaw in their evolution, but because we've been actively killing them. If there is anything we can still do for them, we have to do it, and to the best of our abilities! However, we also need to grow wise to the fact that we should have never let it come this far."

Apart from his expertise in molecular biology and veterinary medicine, Diecke also possesses an artistic streak. He is the co-author of the "Bricolage" artistic installation, which had gone over very well at the Ars Electronica festival in 2021.

In collaboration with two Australian artists, Diecke managed to bring silk to life. The authors of the installation created cardiac muscle cells from a blood drop, only to insert them in pieces of silk cloth. The strange "beings" thus created moved according to the rhythm of a heartbeat. . . . Perhaps prodding the viewer toward the question hinted at by the authors: *Just what is it that makes human beings feel superior to all others?*[28]

Are humans superior to the northern white rhinos? Our answer would be that what superiority we may possess has long been reduced to the sheer ingeniousness of our methods when it comes to slaughter. If a boundless aptitude for cruelty represents the pinnacle of the human intellect, the planet would have been much better settled by the "not-so-bright" northern white rhino.

"Perhaps what I'm about to say won't be considered ethically correct . . ." Diecke allowed himself to observe. "But we've spent so much money on saving individual humans on an overcrowded planet, and so little on saving those animals headed for extinction!"

He was of course the first to agree that such ponderings can be very dangerous, especially because of all the crazy ideas they could lead to. "Well," he said with a tired smile, "all I can do to maintain my own equilibrium is to keep helping both humans and animals."

When it comes to the project of saving the northern white rhino, Diecke is especially pleased with the fact that the fruits of all his past work are finally becoming tangible. "The things I do with stem cells cannot be seen with the naked eye, since the lab events are so tiny as to be almost infinitely remote from our normal experience. With the Bio-Rescue project, however, everything is perfectly clear. We know what our mission is. We know what we can do. And we know that our actions will eventually lead to a baby rhino. For me, this understanding we all share is the beauty of our project."

The key prerequisite for the application of these technologies is access to vital cell material. The cryobanks, the libraries of life, enable the said material's storage in liquid nitrogen at minus 196 degrees Celsius. At this temperature, most of the biological processes are stopped, while cell information is preserved for a few decades or perhaps even centuries.

Frozen Zoo is the world's first cryocollection of endangered animal species, set up by the Beckman Center for Conservation Research at the San Diego Zoo.

The collection was created by the German American pathologist and geneticist Kurt Benirschke in 1975. When he was starting out, most of the technologies for utilizing the collection's contents were not yet available. But Benirschke felt we were nonetheless obliged to collect things "for reasons we might not yet understand."

Frozen Zoo was thus constantly updated. Today it is the largest collection of its kind in the world. Its vaults contain over 10,000 cell cultures, oocytes, sperm samples, and embryos of more than a thousand endangered animal species—and of a single extinct one, the Hawaiian po'ouli bird (*Melamprosops phaeosoma*).

Substantial cryo-collections can also be found at the Smithsonian Conservation Biology Institute, the Haute-Touche natural reserve in France, and at the Leibniz-IZW institute in Berlin. There are also several specialized cryobanks, like Amphibian Ark (specializing in amphibian animals), the ZooParc de Beauval collection (specializing in elephant sperm), etc . . .

The material which could prove instrumental in saving the northern white rhinos is stored at several locations in several different countries. It consists of 300 milliliters of sperm donated by five deceased northern-white males, and the fibroblasts (connective tissue cells) of twelve northern white rhinos. Eight of the twelve animals were not related to each other. Five of them never reproduced. The fibroblasts were obtained through biopsies of rhino skin, performed over the 1976–2016 period.

Along with Najin and Fatu, this precious working material is one of the foundations for the species' survival.

For a more detailed explanation, Sebastian Diecke turned the floor to his colleague Vera Zywitza, a molecular biologist who performs most of the lab work.

Zywitza warmly greeted us from her own Zoom window and the backdrop of her airy bright home. "We're working with the frozen fibroblasts of deceased northern white rhinos," she began carefully, as if addressing a pair of schoolchildren. "We of course first classified the cells and checked them for reproductive viability and genetical soundness. Then we began reprogramming them into induced pluripotent stem cells."

In March 2022, *Nature* magazine published an article co-authored by Vera Zywitza, stating that two teams lead by Sebastian Diecke and Professor Micha Drukker achieved a major breakthrough on the path to the creation of northern-white-rhino oocytes with the help of stem cells.[9]

From Nabire's skin cells, Professor Drukker and his team created induced pluripotent stem cells through the method of episomal reprogramming. Molecules of foreign DNA, or plasmids, were inserted into the fibroblast genomes. This marked the first time scientists managed to create induced pluripotent stem cells from the skin samples of a rhino of such an advanced age.

Yet this remarkable feat also suffered from a significant flaw.

If scientists were to use the induced pluripotent stem cells to create reproductive cells, they would run a huge risk of pathological mutation. The laboratory mice exhibited a marked tendency for tumor development. And so, these cells are currently considered inappropriate for use in assisted-reproduction procedures.

Sebastian Diecke's research team, however, managed to overcome the fatal flaw. Instead of plasmids with their foreign DNA, Diecke's team decided to use RNA viruses. The induced pluripotent stem cells thus created contain no harmful additions and have therefore been found safe for further use.

"We've already managed to create functional cardiac muscle cells, nervous cells and cells of internal organs," Vera Zywitza described the recent advances.

The next step is to use induced pluripotent stem cells to create primordial germ cells—immature reproductive cells, or the precursors of eggs and sperm to germ cells.

"As a matter of fact, we've already managed to do so—in close collaboration with our Japanese colleagues," Zywitza reported. "But the precursor cells are still at a stage when we can't tell if they are male or female. To create oocytes from them, an "environmental niche" is needed—meaning the ovaries. Since we are unable to obtain these organs from the existing northern white rhinos, we first have to create new stem cells in order to create ovary tissue."

The jump from precursors of oocytes (eggs) to fully functional oocytes will be filled with challenges, Vera Zywitza warned us.

"When precursors of oocytes are available," she predicted, "we will be able to create oocytes, perform artificial fertilization, transfer the embryos into surrogate southern-white mothers—and then we will finally have northern-white babies. If we manage to do it in time, at least the first ones will be able to grow up with Najin and Fatu. The hope is that the two of them will be able to transmit the species' cultural heritage to the next generation. Which means we need to hurry."

Zywitza's impatience could hardly be held against her, given the ferocity of the race against time the BioRescue team is facing. And especially since, in all of Germany, only two pairs of hands are currently involved in stem-cell research within the BioRescue project: Zywitza and her technical assistant.

"We would like to do more, but we would need at least six more hands," Diecke weighed in somberly.

His colleague Zywitza did not let it go at that. "I can't understand how more money can be spent on the material than on personnel," she frowned. "Since science is such a repetitive process, we need more staff! You really have to do things over and over and over again. Which means that just two people can't really achieve that much."

Even with all the high-tech contraptions available, the process of discovery still entails huge amounts of slow and laborious manual work. It bears mentioning every single induced pluripotent stem cell has to be created "by hand." To surmount this obstacle, several start-up companies have sprung up, trying to set up the mass production and differentiation of induced pluripotent stem cells.[10]

These companies' chief mission, of course, is to reap the rewards of the utilization of stem cells in medical treatments.

The second great challenge is posed by the standardization of automated production. In April 2021, the French company TreeFrog Therapeutics announced its ability to produce fifteen billion induced pluripotent stem cells in a week. All the resultant therapies, however, still have to be authorized before they can be used for medical treatment.

🦏

"It took us three years to develop primordial germ cells. My guess, which could prove completely wrong, is that we'll need three to six years to create oocytes—given, of course, adequate financial backing and sufficient staff. Otherwise, we will not succeed at all," Sebastian Diecke estimated.

The "friendly rivals" from the San Diego Zoo are also trying to create northern-white egg cells by reprogramming skin cells of deceased animals. In the past, the American scientists managed to pull slightly ahead of the European-Japanese partnership. In 2011, they created induced pluripotent stem cells of two endangered animal species, stored in the Frozen Zoo cryo-collection: the northern white rhino and the drill (*Mandrillus leucophaeus*).[11]

Their procedure suffered from the same shortcomings as Micha Drukker's procedure. To trigger the reprogramming of skin cells, they inserted foreign genes, which can get integrated into the recipient's genome and trigger pathological changes.

In 2021, the team lead by Jeanne Loring, a stem-cell researcher, and the founding director of the Center for Regenerative Medicine at the Scripps Research Institute, managed to create induced pluripotent stem cells with no unwanted additions. As Loring told *Nature* magazine in 2021, her team had already reached the stage of trying to create precursor reproductive cells from induced pluripotent stem cells—following the example of Katsuhiko Hayashi, who created egg cells from the skin cells of mice.

"We are now trying to find an efficient way to make the precursors of gametes, known as primordial germ cells, from the induced pluripotent stem cells of northern white rhinos. We know it's possible—we've seen it happen spontaneously in cultures of these induced pluripotent stem cells—but we need to learn how to generate more of them. And then we have to turn those germ cells into eggs and sperm—or at least, something like sperm," Loring told *Nature* magazine.[12]

With some bitterness in his voice, Sebastian Diecke ventured the opinion that the level of collaboration with the American scientists could be better.

"In science, those who arrive first get all the attention and financing," he shook his head. "The San Diego Zoo team has been researching stem-cell technology for a long time. They also set up the first endangered species cryobank, so they were able to attract lot of interest. On the other hand, this put them under a great deal of pressure, since then their private donors are always demanding results."

Diecke nonetheless believes that the San Diego operation lacks an expert of Hayashi's stature.

"Only a handful of people possess comparable experience in the field," he ventured. "And as far as I can tell, the San Diego team is collaborating with none of them. They are more focused on creating as many induced pluripotent stem cells as possible—and on their characterization on the molecular level. They are also trying to understand the northern-white-rhino genome. This enables us to collaborate without directly competing on specific research projects. But given all the data we're creating, our team could actually prove faster in the long run."

There is another obstacle preventing the collaboration between the European-Japanese and the American projects—an obstacle closely rhyming with human nature. The fact is that the San Diego Zoo refuses to share its cryobank material—the fibroblasts of twelve thoroughbred northern white rhinos—with the BioRescue researchers.

In her lab, Vera Zywitza thus only has access to the fibroblasts of five northern white rhinos and one southern-northern hybrid.

"If we want to save the species, we must collaborate. I think the scientists from San Diego realize that. We sometimes share information on the protocols—what works for us, what works for them—but not material," Zywitza was clear.

Diecke further delineated the complex relationship: "Overall, we try to keep each other posted. We exchange information from papers and articles which are about to be published. Unfortunately, no coordinated meetings take place. We would very much like that to change. If we could agree on one team focusing on oocyte creation, and the other on sperm cell creation, the solutions would be found much faster."

The German scientist went on to add there were several levels of cooperation between the two camps. On the level of management—all too often involving poster-boys and poster-girls who all too often resemble politicians—there is a lot more competitiveness. The ones doing the actual work are much more inclined to collaborate.

Cesare Galli, the "father" of northern-white-rhino embryos, firmly believes stem cells are the path to take.

"They represent the future of rescuing endangered species," he told us. "Though it's also true these technologies still have a long way to go. It's one thing to create the egg cell of a mouse, and quite another to create the egg cell of a much larger animal like the northern white rhino. Or a human being. True, we've made a lot of progress. But we're still quite far from the moment when we'll be able to create the first rhino embryo from stem cells."

According to Galli's estimate, we could see the first offspring on the basis of stem cells in ten to twenty years. "But why shouldn't we try?" he called out passionately. "Oliver Ryder, a geneticist with the San Diego Zoo, wrote a scientific article stating the genetic diversity of the stored cell and tissue samples is greater than that of the entire current population of the southern white rhinos, comprising some 20,000 specimens."

Galli is perfectly ready to acknowledge cloning as the third available option for rescuing the northern white rhinos. "But so far, the conservationists haven't been able to accept cloning as a viable alternative. I couldn't really tell you why. We even had problems with explaining the need for assisted reproduction. They told us we were wasting time and money. . . ."

Up to this point, the BioRescue scientists hadn't yet discussed cloning, Galli admitted. Yet in his professional opinion, the next generation of northern white rhinos will be produced faster through cloning than through stem-cell technologies.

"In the short run, cloning is the better solution. Not in the long run, no—but we need to get the first surrogate mother pregnant as soon as possible. Focus, people, focus—right?"

Resurrecting extinct animals is no longer the stuff of science-fiction literature; it is fast becoming reality.

Several research projects are taking place, aimed at bringing back extinct species or enhancing the genetic diversity of critically endangered species. The methods employed range from stem-cell technologies and genomics to cloning and assisted reproduction techniques. The available technology grows ever more precise, efficient, fast, and cheap.

"Biotechnologies, like cloning and gene editing, now give us a chance to accelerate the evolution of species so they can actually cope with change and survive it," claims Ben Novak in Mongabay's article. Novak is a lead scientist with the *Revive & Restore* American non-profit organization striving for the enhancement of biodiversity through the genetic rescue of critically endangered and extinct species.[13]

In 2009 *Revive & Restore* cloned the Pyrenean ibex, extinct from 2000 on. The clone survived for only a few minutes. The next two experiments, however, proved much more successful. In 2020, the non-profit organization cloned a Przewalski's horse it decided to name Kurt. In December of the same year, the first black-footed ferret (*Mustela nigripes*) was cloned—a female named Elizabeth Ann.[14]

Both the Przewalski's horse and the black-footed ferret belong to the unfortunate category of endangered species.

The Przewalski's horse went extinct in the wild in the sixties, though the species was fortunately preserved through a breeding program, from which a number of these animals were released back into the Mongolian steppe.

The greatest threat to their survival is currently posed by their low genetic diversity. All 2,000 of the horses are descended from twelve specimens saved from extinction around 1900. Kurt was cloned from the cellular material of his deceased predecessors stored in liquid nitrogen at San Diego's Frozen Zoo. In five to ten years, when the young male enters his reproductive age, the hope is that he can make a significant contribution to the genetic refreshment of his species.

The story of the black-footed ferret, one of Africa's most critically endangered animals, is depressingly similar. All existing member of the species—some 650 of them—are descended from only seven animals, so inbreeding is a significant problem.

By cloning Elizabeth Ann, the researchers were determined to enhance the population's genetic diversity. Elizabeth Ann was created from the genetic information of a female black-footed ferret, stored in liquid nitrogen from 1988 on. The said female's genome differed significantly from that of all the currently living black-footed ferrets, so hopes for an important boost to the species' genetic diversity are running high.

In February 2022 Elizabeth Ann turned one year old. By ferret count, this meant she was ready to reproduce. The researchers are still looking for a suitable male—above all a sufficiently gentle one. The mating of the black-footed ferrets can turn rather rough, and Elizabeth Ann is a highly precious commodity.[15]

The cloning of both species triggered a tremendous backlash. A number of highly critical voices were raised, demanding what it was all for.

The scientists provided a well-argued response, based on the ethical analysis of cloning as a means to the genetic rescue of the black-footed ferret. Performed beforehand, the analysis ticked off all the right boxes: were the goals justified, could cloning be performed in a sufficiently responsible fashion, can the project count on public support?

With the black-footed ferrets, cloning was indeed proved to be the sole means of salvation. The remaining ferrets suffered from high rates of inbreeding, and there was no other specimen available to boost the overall genetic diversity. This was how the facts were presented to the *Mongabay* webpage by Samantha Wisely, a conservation geneticist with the

University of Florida and a member of the team which had performed the ethical analysis.

Wisely also stressed the need for weighing the soundness of cloning as a means of genetic rescue on a purely case-by-case basis.

A lot of media attention was recently stirred by the American bio-tech start-up Colossal's announcement it aimed to revive the wooly mammoths—or, more precisely, genetically modified Asian elephants adapted to Arctic conditions.

Colossal, founded by entrepreneur Ben Lamm and Harvard genet-icist George Church, cites the fight against climate change as a viable reason for resurrecting the long-gone species.[16]

The wooly mammoths—enormous beasts whose pelt hairs reached up to a meter in length—populated a great part of the Northern hemi-sphere some 50,000 years ago. Most of them died out after the end of the last ice age approximately 12,000 years ago, with the last of them perishing in Siberia some 2,000 years later.

The reason for their extinction could be either mankind or climate change after the ice age's end. Or, of course, both.

Scientists now intend to create woolly-mammoth approximations with the help of the "genetic scissors" method (CRISPR-Cas9), which enables the modification of living organisms through precise cuts into the genome for the insertion of desired genes. Key mammoth genes—like the ones coding for lengthy fur, additional adipose tissue, short ears, the release of oxygen into the bloodstream at low temperatures—will be inserted into the genome of the mammoth's closest relative, the Asian elephant.

Colossal believes the result should be a highly cold-resistant Asian elephant, possessing all the biological characteristics of the woolly mammoth.

The genetic material was obtained from ancient mammoth tusks, found amid the melting permafrost. The genome was reconstructed with the help of the latest DNA sequencing technologies. The plan is to insert the important mammoth genes into an elephant egg cell.

Since the Asian elephants are endangered, the scientists will forego using their egg cells, but will rather create fresh ones through reprogramming Asian-elephant skin cells.

The embryos predicted to result from the process are planned to mature inside an artificial elephant womb, which the scientists aim to develop.

Many of these procedures have never been performed on elephants. Several of the required technologies have not yet even been invented. However, the Colossal company already raised $15 million to fund the project. The first baby mammoth is projected to be born in five years.

It was also announced that the animals will be settled at the Pleistocene Park in northeastern Siberia, founded by Sergey and Nikita Zimov.

The father-and-son scientific team has settled a fenced-off part of the tundra near the Chersky urban locality with elks, reindeers, bison, Bactrian camels, and other large herbivores. The aim was to observe the animals' influence on landscape transformation. According to the Zimovs' predictions, an influx of a sufficient number of "mammoths" and other herbivores should help transform the Arctic tundra into grass-covered plains.

Permafrost is currently melting under the tundra, threatening to release enormous quantities of methane into the atmosphere. The process would be slower if the ground was covered with grass, the Zimovs have posited. The grass, made up of lighter hues than the tundra's shrubs and trees, should soak in less heat. And the animals' constant trampling of the surface should additionally slow down the methane leakage, buying us more time to prevent or mitigate the impending disaster.

Talk of resurrecting the mammoth is a periodically recurring phenomenon, we were warned by Victoria Herridge, a paleontologist at London's Natural History Museum and a prominent critic of Colossal's project.[17]

Herridge had turned down the offer of a seat on the start-up's advisory board—not because she questioned Colossal's motives, but on account of her own ethical considerations. "The reason why I said no, was really more at me than at them," Herridge told us over Zoom. "I felt I wanted to retain my ability to be an independent voice. I didn't feel that I could do that even though I had assurances they were interested in independent and critical thought," she explained.

In the London-based paleontologist's view, the transformation of our planet shouldn't be left to an elite few. While deciding on future research projects, companies like Colossal should actively consult the public. "The most important thing about this whole debate is that it is pushed out into the open and pushed out toward the public. It's really important that the public is asked about its opinion on de-extinction projects, while them being in the hands of a small number of people and private funders with decisions taken behind closed doors," Herridge stressed.

She doubts the Colossal's argument of ecosystem restoration and climate change mitigation as a justification for its research. "Even if we get a hundred thousand mammoth-like animals there I'm unconvinced of the timeframe that change will happen and whether it could be going in a way that is meaningful to counteract global warming. That's even without considering the aspects of what role did mammoths really play, that we still don't know," Herridge told us over Zoom.

Not only Colossal intends to do experimentation on animals but also on ecosystems. "The point of the plan is to have a genetically modified Asian elephant that could live in Siberia, and their reason to put it in there is to modify the ecosystem. If you are going to modify an ecosystem, you are going to think very carefully about the multi-faceted aspects of the ethics of that. How is it going to impact people, the climate.... If you are trying to geo-engineer the climate, you are going to change the whole world. I wonder if this is a good thing. That's why it's important to have straight ethical procedures—open and clear-cut; decided by independent people," Herridge said.

As for the wooly-mammoth project, she believes we would do much better to focus on saving the Asian elephants than on pondering how Asian elephants could be transplanted to the Arctic. The same goes for the elephants' natural habitats, which, after all, are key for the species' survival. "I don't really know where we end up, if we'll have a lot of elephants in Siberia. And, for example, no Asian elephants in the ecosystems where they are the keystone species. We are losing what we want to conserve, while we are creating something new. This is pushing the discussion to whole different territory—it's a question in what kind of world we want to live," she continued.

"If I were to advocate to myself, the strongest argument would be to expand the potential range in which Asian elephants could live. As this genetically-modified elephant—basically mostly Asian elephant, which is a highly threatened species—could expand the range in which it could live on the planet, then you could increase the likelihood that an elephant of some kind at least genetically exists for longer," Herridge reasoned.

But a species is more than just genes, she exclaims. "It's part of an ecosystem and conservation. It's about looking after the planet and preserving ecosystems."

On the other hand, Herridge shared her enthusiasm about the work of Thomas Hildebrandt on rhino and elephant reproduction, and the project of saving the northern white rhinos. There could be much more direct benefit to other rhino species than for an Asian elephant in the wooly-mammoth project, she believes.

She has underlined the unique ethical risk assessment procedure of the BioRescue project. "They are weighing up the interest of one individual against the interest of a species. This is really an interesting ethical question. Does one individual trump the ethical rights of future potential individuals of northern white rhinos? Should she suffer if it means that in the future there would be potentially a million of happy northern white rhino individuals? It's great that they ask themselves these questions. But this certainly slows you down, and it can be a big problem, if you are in hurry."

When reading about the plan for the resurrection of hybrid mammoths, or restoration of the dodo bird, which is the latest plan of Colossal, the ever more popular word *frankenscience* quickly springs to mind. But saving the northern white rhinos through advanced biomedicine has nothing to do with that, Sebastian Diecke vehemently shook his head.

"We're not tinkering with mammoths or bringing a neanderthal back to life," he smiled disarmingly. "We're only trying to right a recent wrong.

We're not meddling with the rhinos' genome. We are doing everything in our power to get things right. Therefore, we're so concerned with monitoring the cells' quality. And also, why we appointed an ethical-risk committee. . . . So that all the project's aspects can be assessed, not only the scientific ones."

Was he at all fascinated by the seventy-million-year-old fossilized dinosaur embryo found inside an egg primed to hatch only a few days before the embryo's death? The media covered the find of the marvelously preserved embryo—the perfect miniature of a grown dinosaur—a few days before we talked to Diecke. Was he not at least tempted to turn his tremendous expertise to helping a baby dinosaur see the light of day?

"That was Jurassic Park, and this is BioRescue. Two completely different things," Diecke grimaced before we could fully form the question.

"There is no more space on the planet for the dinosaurs," he elaborated. "We are unable to provide them with the environment they require. If we brought them back, it would be for entertainment purposes. And you also mustn't forget that we don't have any stores of their vital cell material. We can't do everything on the basis of mere genetic information. We may one day be able to modify the tissue of other animals to recreate the entire genome of a long-vanished species, but. . . ." Diecke opted to conclude the sentence with an eloquent shrug.

"We're not playing God, we're only supporting nature," Vera Zywitza chipped in. "We're helping the rhinos with what they can no longer do on their own."

Neither Diecke nor Zywitza have yet visited "the girls" at Ol Pejeta. "I would so very much like to meet them," Zywitza told us at the start of 2022. "But I believe I am far more useful in the lab than out in the field."

Diecke was in full agreement with his colleague. "We decided not to waste any funding on a trip to Kenya, since we really couldn't do all that much there. Which of course doesn't mean that I, too, am not longing to meet Najin and Fatu. I simply adore animals. We did visit the Dvůr Králové Zoo, where we got the chance to pet southern white rhinos. They may not be the smartest creatures on the planet, but they are still mesmerizingly wonderful. They look ancient—they even slightly reminded me of dinosaurs. . . ."

Diecke trailed off. Then he quickly changed the subject—as if to fend off the thought that the northern white rhinos and the dinosaurs also seem bound to share the same fate.

The only difference being that the asteroid has now all but been replaced by *homo sapiens*.

Epilogue

In pushing other species to extinction, humanity is busy sawing off the
limb on which it perches.
        —Paul Ehrlich, *The Population Bomb*[1]

The majority of known species went extinct over the last
three hundred years.

The majority of them over the last fifty years.

The majority of those over the last ten years.

Here and now, roughly 1 million plant and animal species are teth-
ering on the brink of extinction .[2]

In this context, saving the northern white rhino is certainly a worthwhile
project.

And a very necessary one.

The project's eventual fruition would not merely save an essentially
extinct species. Through its newly developed technologies, experience,
and infrastructure, it could pave the way to the salvation of numerous
other critically endangered animals. Perhaps even some of those whose
tragic fate currently garners much less public sympathy than the fate of
more iconic species.

Of course, that is also the main reason why the BioRescue project must succeed. Ecosystems and their biodiversity simply must be
preserved. If we are at all interested in the survival of our own species,
we have to do everything in our power to deter the so-called domino

effect—the devastating chain reaction triggered by the northern white rhino's permanent extinction.

So let us suppose that, if all goes well, the first northern white rhino after Fatu will be born in 2024. Then let us summon even more optimism and suppose that the stem-cell technologies will be able to boost the northern white's genetic pool to the extent that we can one day claim to have rescued a species we had almost exterminated.

Let us suppose all that. But what kind of world will the new northern white rhinos be born into? Will they, as seems highly likely, be forced to live inside savagely protected refugee camps? Are more-or-less luxurious prisons the best we can offer them?

The last of the wilds are vanishing as we speak.

The story of the northern white rhino's extinction is the story of war, racism, climate change, and socio-economic Mariana trenches. In a phrase, it is the story of the governing primates' vanity; it is a parable of the all-devouring Anthropocene.

When rescuing any single species, the brunt of our efforts should be directed towards eliminating the main factors facilitating its extinction. Just like with our desperate fight against climate change, it will be impossible to win without a series of paradigm shifts.

How highly would you rate the chances of northern white rhino offspring born into a world that had already once exterminated their species?

But then again, what is accomplished if we simply give up in advance and we allow ourselves to be swept along in the general tide of cynicism and apathy?

"I worry that so-called 'de-extinction' lulls everyone into a false sense of security. No one is going to care if they (wrongly) believe that extinction is just a temporary inconvenience, and all it takes is a bit of technology to bring a species back. It doesn't work like that," the British zoologist Mark Carwardine wrote in the epilogue of the new edition of *Last Chance to See*, more than thirty years after he and Douglas Adams embarked on their search for the most endangered species.[3]

Carwardine also pointed out that, during his lifetime, 25 percent of all rhinos were killed.

In 1988, when he and Adams set out on their quest, the planet was home to five billion human beings. Now, in the spring of 2023, the number has risen to eight billion. Since 1970, when there were four billion less of us, we lost more than a half of all wild animals.[4]

Eight out of ten countries with the youngest population are located on the African continent. The world's youngest country is Niger, where the average age has been estimated at 14.8 years. Niger is located in the middle of the Sahel, one of the planet's hottest and driest regions, where the wages of climate change are evident everywhere. It is no coincidence that Niger also serves as a key point of departure for the African migrations to Europe via the Mediterranean.[5]

The average age in Kenya, where Najin and Fatu currently reside, is 17.1 years.

"We haven't solved any of the original problems. The forces that whipped out the northern white rhino in the wild—war, poverty, and poaching—haven't magically gone away. Despite some spectacular successes, rhino poaching is still at crisis level. So why would we want to bring them back into such an unsafe and hostile world," demanded Carwardine, who devoted his entire life to the conservation of critically endangered species.[6]

He also managed to come up with an answer.

According to him, the northern white rhino and other critically endangered species are worth saving because without them, the world would be an even darker and lonelier place.

We couldn't agree more.

Notes

Epigraph

1. Douglas Adams and Mark Carwadine, *Last Chance to See* (New York: Ballantine Books, 1992), London.

Introduction

1. David Attenborough with Jonnie Hughes, *A Life on Our Planet: My Witness Statement and a Vision for the Future* (New York: Grand Central Publishing, 2020)
2. Amitav Ghosh, *The Nutmeg's Curse* (Chicago: University of Chicago Press, 2021), 204.

Chapter One

1. The International Journal of Conservation, "A last chance to save the northern white rhino?" *Oryx* 2008, http://www.rhinoresourcecenter.com/pdf_files/124/1245681966.pdf (June 2009).
2. Save the Rhino, "The last northern white rhinos," https://www.savetherhino.org /rhino-species/white-rhino/the-last-northern-white-rhinos/ (19 Nov. 2014).
3. Save the Rhino, "The last northern white rhinos," https://www.savetherhino.org /rhino-species/white-rhino/the-last-northern-white-rhinos/ (19 Nov. 2014).
4. BBC, "Rwanda genocide: 100 days of slaughter," https://www.bbc.com/news /world-africa-26875506 (4 Apr. 2019).
5. National Library of Medicine, "Continent-wide survey reveals massive decline in African savannah elephants," https://www.ncbi.nlm.nih.gov/pmc/articles/PMC5012305 / (31 Aug. 2016).
6. Curry Lindhal, "War and the white rhinos," *Rhino Resource Center*, 1972, http://www.rhinoresourcecenter.com/index.php?s=1&act=refs&CODE=note_detail &id=1165245054 (Oct. 2022).
7. Kes Hillman Smith, *Garamba: Conservation in Peace & War* (self-publishing, 2014), 243–49.
8. Kes Hillman Smith, *Garamba: Conservation in Peace & War* (self-publishing, 2014), 211–24.
9. Eastern Congo Initiative, "History of the conflict," https://www.easterncongo.org /about-drc/history-of-the-conflict/ (Oct. 2022).

10. ReliefWeb, "Democratic Republic of Congo Update," https://reliefweb.int/report /democratic-republic-congo/democratic-republic-congo-update-dec-2000 (31 Dec. 2000).

11. Douglas Adams, Mark Carwardine, *Last Chance to See* (Arrow Press, 2020, new edition), 90.

CHAPTER TWO

1. Channing Sargent, "African black rhino population increasing," *One Earth*, https:// www.oneearth.org/african-black-rhino-populations-increasing/ (25 Oct. 2021).

2. Ol Pejeta Conservancy, "Annual report 2021," https://issuu.com/olpejetaconser vancy/docs/ol_pejeta_conservancy_2021_annual_report_final_dra (30 Aug. 2022).

3. UN, "Severe drought threatens 13 million with hunger in Horn of Africa," https:// news.un.org/en/story/2022/02/1111472 (8 Feb. 2022).

4. World Bank, "Climate Migration in Africa: How to Turn the Tide," https://www .worldbank.org/en/region/afr/publication/climate-migration-in-africa-how-to-turn -the-tide (5 Jan. 2022).

5. United Nations, "Africa is particularly vulnerable to the expected impacts of global warming," https://unfccc.int/files/press/backgrounders/application/pdf/factsheet_africa .pdf (Oct. 2022).

6. PBS, "Kenya's worst drought in decades creates humanitarian crisis," https://www .pbs.org/newshour/show/kenyas-worst-drought-in-decades-creates-humanitarian-crisis (14 Jan. 2022).

7. AP, "Hundreds of elephants, zebras die as Kenya weathers drought," *WNYT* 2022, https://wnyt.com/associated-press/international/hundreds-of-elephants-zebras-die-as -kenya-weathers-drought/ (4 Nov. 2022).

8. Medium, "How the vulture and the little girl ultimately led to the death of Kevin Carter," https://medium.com/@denislesak/how-the-vulture-and-the-little-girl-ultimate ly-led-to-the-death-of-kevin-carter-d9871c4137f2 (18 Dec. 2015).

9. Business Daily Africa, "Mt Kenya glaciers to disappear by 2040," https://www .businessdailyafrica.com/bd/economy/mt-kenya-glaciers-to-disappear-by-2040 -3588496 (Oct. 2021).

10. NCBC News, "Fighting for Land and Water," https://www.cbc.ca/newsinterac tives/features/fighting-for-land-and-water (18 Dec. 2021).

CHAPTER THREE

1. Acams Today, "Europe and the Illegal Wildlife Trade: Where Are We Now?," https://www.acamstoday.org/europe-and-the-illegal-wildlife-trade-where-are-we-now / (22 Feb. 2022).

2. Vanda Felbab-Brown, *The Extinction Market: Wildlife Trafficking and How to Counter It* (Oxford University Press, 2017), 1.

3. ICE, "End wildlife trafficking," https://www.ice.gov/features/wildlife (20 Sep. 2022).

4. Rachel Love Nuwer, *Poached: Inside the Dark World of Wildlife Trafficking* (Scribe, 2018), 37–39.

5. Rachel Love Nuwer, *Poached: Inside the Dark World of Wildlife Trafficking* (Scribe, 2018), 55–56

6. Kees Rookmaaker, "Distribution and extinction of the rhinoceros in China: review of recent Chinese publications," *Pachyderm* 2006, https://www.researchgate.net/publica tion/285764437_Distribution_and_extinction_of_the_rhinoceros_in_China_review_of _recent_Chinese_publications (Jan. 2006).

7. Save the Rhino, "What's the verdict for rhinos?," https://www.savetherhino.org /africa/cites-18th-cop-whats-the-verdict-for-rhinos/ (27 Aug. 2019).

8. RFI, "Asian consumers, criminal gangs drive rhino poaching in Africa," https:// www.rfi.fr/en/africa/20190319-asian-consumers-criminal-gangs-drives-african-rhino -poaching (20 Mar. 2019).

9. Rachel Love Nuwer, *Poached: Inside the Dark World of Wildlife Trafficking* (Scribe, 2018), 171–85.

10. Julian Rademeyer, *Killing for Profit: Exposing the Illegal Rhino Horn Trade* (Zebra Press, 2012), 102.

11. Julian Rademeyer, *Killing for Profit: Exposing the Illegal Rhino Horn Trade* (Zebra Press, 2012), 286.

12. CITES,"Rhinoceroces,"https://cites.org/eng/prog/terrestrial_fauna/Rhinoceroces, (Nov. 2022).

13. Julian Rademeyer, *Killing for Profit: Exposing the Illegal Rhino Horn Trade* (Zebra Press, 2012), 22.

14. Vanda Felbab-Brown, *The Extinction Market: Wildlife Trafficking and How to Counter It* (Oxford University Press, 2017), 113

15. Keith Somerville, *Ivory: Power and Poaching in Africa* (Hurst & Company, 2016), 9–10.

16. Keith Somerville, *Ivory: Power and Poaching in Africa*, (Hurst & Company, 2016), 34.

17. Keith Somerville, *Ivory: Power and Poaching in Africa*, (Hurst & Company, 2016), 107–11.

18. Keith Somerville, *Ivory: Power and Poaching in Africa*, (Hurst & Company, 2016), 123.

19. University of Massachusetts, "The Great Elephant Census Reports Massive Loss of African Savannah Elephants," https://elephantswithoutborders.org/projects/great -elephant-census/ (1 Sep. 2016).

20. BBC, "China's ban on ivory trade comes into force," https://www.bbc.com/news /world-asia-china-42532017 (1 Jan. 2018).

21. Science Direct, "Understanding determinants of the intention to buy rhino horn in Vietnam through the Theory of Planned Behaviour and the Theory of Interpersonal Behaviour," https://www.sciencedirect.com/science/article/pii/S0921800922000234 (May 2022).

CHAPTER FOUR

1. IZW Berlin, "Captive breeding in giant pandas," https://www.izw-berlin.de/en /captive-breeding-in-giant-pandas-bridging-between-innovative-art-and-reproductive -biology.html.

CHAPTER FIVE

1. Shaoni Bhattacharya, "World's First Cloned Horse is Born," *New Scientist*, 6 August 2003, https://www.newscientist.com/article/dn4026-worlds-first-cloned-horse-is-born / (20 Sep. 2022).

2. European Parliament, "Ban Not Just Animal Cloning, but Cloned Food, Feed and Imports Too, Say MEPs," 17 Jun. 2015, https://www.europarl.europa.eu/news/en/press -room/20150617IPR67269/ban-not-just-animal-cloning-but-cloned-food-feed-and -imports-too-say-meps (20 Sep. 2022).

3. Dr. Martin Wilding, "What Is the Difference Between IVF & ICSI?," *ABCIVF*, 3. Jan. 2019, https://www.abcivf.co.uk/blog/what-is-the-difference-between-icsi-and-ivf (20 Sep. 2022).

4. Dolly the Sheep @20, "The World's Most Famous Sheep. 20 Years on From Her Birth We Examine Her Lasting Legacy," https://dolly.roslin.ed.ac.uk/index.html (20 Sep. 2022).

5. The Noble Prize, "Sir Martin J. Evans—Facts," https://www.nobelprize.org/prizes /medicine/2007/evans/facts/ (19 Sep. 2022).

6. Bruno Simini, "Cremona Italian Scientist Investigated After Animal Cloning Experiment," *The Lancet*, 16 Oct. 1999, https://www.thelancet.com/journals/lancet/article /PIIS0140-6736(05)76221-8/fulltext (20 Sep. 2022).

7. BBC News, "Italian Laboratory Clones 14 Pigs," 28 Oct. 2005, http://news.bbc .co.uk/2/hi/science/nature/4386510.stm (20 Sep. 2022).

8. Simar Bajaj, "Why Did the First Human Patient to Receive a Pig Heart Trans-plant Die?," *Smithsonian Magazine*, 14 July 2022, https://www.smithsonianmag.com /science-nature/why-exactly-did-the-first-human-patient-to-receive-a-pig-heart -die-180980361/ (20 Sep. 2022).

9. Sophie Kevani, "US FDA Declares Genetically Modified Pork 'Safe to Eat,'" *The Guardian*, 17 Dec. 2020, https://www.theguardian.com/environment/2020/dec/17/us-fda -declares-genetically-modified-pork-safe-to-eat (20 Sep. 2022).

CHAPTER SIX

1. AboutZoos, "Dvur Kralove Zoo," https://aboutzoos.info/zoos/zoo-database /europe-zoo-database/168-dvur-kralove-zoo.

2. https://www.youtube.com/watch?v=wZYe3LqN8q0.

CHAPTER SEVEN

1. World Organisation for Animal Health, "Our Definition of Animal Welfare," https://www.woah.org/en/what-we-do/animal-health-and-welfare/animal-welfare/ (20 Sep. 2022).

2. National Centre for the Replacement, Refinement & Reduction of Animals in Research, "What are the 3Rs?," https://www.nc3rs.org.uk/who-we-are/3rs (20 Sep. 2022).

3. Multidisciplinary Digital Publishing Institute, "An Ethical Assessment Tool (ETHAS) to Evaluate the Application of Assisted Reproductive Technologies in Mam-

mals' Conservation: The Case of the Northern White Rhinoceros (*Ceratotherium simum cottoni*)," Animals | Free Full-Text | An Ethical Assessment Tool (ETHAS) to Evaluate the Application of Assisted Reproductive Technologies in Mammals' Conservation: The Case of the Northern White Rhinoceros (Ceratotherium simum cottoni) (mdpi.com) (20 Sep. 2022).

4. National Library of Medicine, "Scientific and Ethical Issues in Exporting Welfare Findings to Different Animal Subpopulations: The Case of Semi-Captive Elephants Involved in Animal-Visitor Interactions (AVI) in South Africa," https://pubmed.ncbi.nlm.nih.gov/31635075/ (20 Sep. 2022).

5. Erin Udell, "CSU professor and 'father of veterinary medical ethics' Bernie Rollin dead at 78," https://eu.coloradoan.com/story/news/2021/11/23/bernie-rollin-colorado-state-professor-animal-ethics-expert-has-died/8720728002/ (20 Sep. 2022).

Chapter Eight

1. WWF, "What is Human-Wildlife Conflict and Why Is It More than Just a Conservation Concern?," https://www.worldwildlife.org/stories/what-is-human-wildlife-conflict-and-why-is-it-more-than-just-a-conservation-concern.

2. Ecolex, "Wildlife Conservation and Management Act, 2013 (No. 47 of 2013)," https://www.ecolex.org/details/legislation/wildlife-conservation-and-management-act-2013-no-47-of-2013-lex-faoc134375/, (2013).

3. Keith Somerville, *Ivory: Power and Poaching in Africa* (Hurst & Company, 2016), 109.

4. CITES, "Ivory Sales Get the Go-ahead," https://cites.org/eng/news/pr/2008/080716_ivory.shtml (21 Jan. 2021).

5. Save the Elephants, "Samburu Elephant Project," https://www.savetheelephants.org/project/samburu-elephant-project/, (2022).

Chapter Nine

1. Wildlife Angel, "1000 Rangers killed in 10 Years Worldwide," https://wilang.org/en/1000-rangers-killed-in-10-years-worldwide/ (2019).

2. Al Jazeera, "Impossible to Rescue," https://www.aljazeera.com/features/2021/12/1/impossible-to-rescue-how-three-foreigners-died-in-burkina-faso (1 Dec. 2021).

Chapter Ten

1. Leibniz Institute for ZOO and Wildlife Research, "Towards the Next Level of Biobanking," https://www.izw-berlin.de/en/towards-the-next-level-of-biobanking.html (21 Sep. 2022).

2. Richard Black, "Endangered Species Set For Stem Cell Rescue," *BBC News*, 4. September 2011, https://www.bbc.com/news/science-environment-14765186 (29 Sep. 2022).

3. Korody ML, Ford SM, Nguyen TD, Pivaroff CG, Valiente-Alandi I, Peterson SE, Ryder OA, Loring JF, "Rewinding Extinction in the Northern White Rhinoceros: Genetically Diverse Induced Pluripotent Stem Cell Bank for Genetic Rescue," *Stem Cells*

Dev. 2021 Feb; 30(4):177–89. doi: 10.1089/scd.2021.0001. *Epub* 2021 Feb 8. PMID: 33406994; PMCID: PMC7891310.

4. Takahashi K, Yamanaka S., "Induction of Pluripotent Stem Cells from Mouse Embryonic and Adult Fibroblast Cultures by Defined Factors," Cell. 2006 Aug 25; 126(4):663–76. doi: 10.1016/j.cell.2006.07.024. Epub 2006 Aug. 10. PMID: 16904174., Induction of pluripotent stem cells from mouse embryonic and adult fibroblast cultures by defined factors—PubMed (nih.gov) (21 Sep. 2022).

5. Erin McCormick, "Eat Just is Racing to Put 'No-Kill Meat' on Your Plate. Is it Too Good to be True?," *Guardian*, 16 Jun. 2021, https://www.theguardian.com/food/2021 /jun/16/eat-just-no-kill-meat-chicken-josh-tetrick (21 Sep. 2022).

6. Will Bedingfield, "Lab-Grown Tuna Steaks Could Reel in Our Overfishing Problem," *Wired*, 23 March 2021, https://www.wired.co.uk/article/blue-nalu-lab-grown-fish (21 Sep. 2022).

7. Cyranoski, D., "Mouse Eggs Made From Skin Cells in a Dish," *Nature* 538, 301, (October 2016), https://doi.org/10.1038/nature.2016.20817 (21 Sep. 2022).

8. Rachel Stewart, "Works—Bricolage by Nathan Thompson, Guy Ben-Ary, and Sebastian Diecke," *Art the Science's Polyfied Magazine*, 15 Sep. 2021, https://artthescience. com/magazine/2021/09/15/works-bricolage-by-nathan-thompson-guy-ben-ary-and -sebastian-diecke/ (21 Sep. 2022).

9. Zywitza, V., Rusha, E., Shaposhnikov, D. et al., "Naïve-Like Pluripotency to Pave the Way for Saving the Northern White Rhinoceros From Extinction," *Sci Rep* 12, 3100 (2022), https://www.nature.com/articles/s41598-022-07059-w#citeas (21 Sep. 2022).

10. Eric Bender, "Stem-Cell Start-Ups Seek to Crack the Mass-Production Problem," *Nature*, 29 September 2021, https://www.nature.com/articles/d41586-021-02627-y (21 Sep. 2022).

11. Scripps Research Institute, "First Stem Cells From Endangered Species," *Science-Daily*, 5 September 2011, www.sciencedaily.com/releases/2011/09/110904140411.htm (21 Sep. 2022).

12. Julianna Photopoulos, "Why Stem Cells Might Save the Northern White Rhino," *Nature*, 29 Sep. 2021, https://www.nature.com/articles/d41586-021-02626-z (21 Sep. 2022).

13. Caitlin Looby, "For Species on the Very Brink of Extinction, Cloning is a Loaded Last Resort," *Mongabay*, 5 January 2022, https://news.mongabay.com/2022/01/for-spe cies-on-the-very-brink-of-extinction-cloning-is-a-loaded-last-resort/ (21 Sep. 2022).

14. Rasha Aridi, "Scientists Cloned an Endangered Wild Horse Using the Decades-Old Frozen Cells of a Stallion," *Smithsonian Magazine*, 15 Oct. 2020, https://www .smithsonianmag.com/smart-news/save-endangered-wild-horse-species-scientists -cloned-stallion-using-its-decades-old-frozen-cells-180976069/ (21 Sep. 2022).

15. Robin McKie, "Wanted: Virile But Gentle Mate For the World's First Cloned Black-Footed Ferret," *Guardian*, 6 Feb. 2022, https://www.theguardian.com/science /2022/feb/06/wanted-virile-but-gentle-mate-for-the-worlds-first-cloned-black-footed -ferret (21 Sep. 2022).

16. Michael Greshko, "Mammoth-Elephant Hybrids Could be Created Within the Decade. Should They Be?" *National Geographic*, 13 Sep. 2021, https://www.nationalgeo

graphic.com/science/article/mammoth-elephant-hybrids-could-be-created-within-the
-decade-should-they-be (21 Sep. 2022).

17. Victoria Herridge, "Before Making a Mammoth, Ask the Public," *Nature*, 20
October 2021, https://www.nature.com/articles/d41586-021-02844-5 (21 Sep. 2022).

EPILOGUE

1. Paul R. Ehrlich, *The Population Bomb* (Buccaneer Books, 1971).

2. Douglas Adams, Mark Carwardine, *Last Chance to See* (Arrow Press, 2020, new
edition), 233.

3. Adams, Carwardine, *Last Chance to See*, 227.

4. Worldometers, "Current World Population," https://www.worldometers.info
/world-population/ (21Sep. 2022).

5. World Atlas, "30 Countries with the youngest population in the world," https://
www.worldatlas.com/articles/the-youngest-populations-in-the-world.html (2022).

6. Douglas Adams, Mark Carwardine, *Last Chance to See*, 227–28.

Bibliography

Adams, Douglas and Carwardine, Mark. *Last Chance to See: A journey in search of our most precious and endangered animals.* Arrow Books (new edition), London, 2020.

Buscher, Bram and Fletcher, Robert. *The Conservation Revolution: Radical Ideas for Saving Nature Beyond the Anthropocene.* Verso, London, 2020.

Dawson, Ashley. *Extinction: A Radical History.* OR Books, New York, 2016.

Deghan, Alex. *The Snow Leopard Project: And Other Adventures in Warzone Conservation.* PublicAffairs, New York, 2019.

Felbab-Brown, Vanda. *The Extinction Market: Wildlife Trafficking and How to Counter It.* Hurst & Company, London, 2017.

Furrel, Errol. *Lost Animals: Extinction and the Photographic Record.* Bloomsbury, London, 2013.

Gosh, Amitav. *The Nutmeg's Curse: Parables for a Planet in Crisis.* John Murray, London, 2021.

Hanks, John. *Operation Lock and the War in Rhino Poaching.* Penguin Books, Cape Town, 2015.

Hillman, Kes. *Garamba: Conservation in Peace & War.* Self-published, Nairobi, 2014.

Jepson, Paul and Blythe, Cain. *Rewilding: The Radical New Science of Ecological Recovery.* Hot Science, London, 2020.

K. Heissem, Ursula. *Imagining Extinction: The Cultural Meanings of Endangered Species,* The University of Chicago Press, Chicago, 2016.

Kolbert, Elizabeth: *The Sixth Extinction: An Unnatural History,* Henry Holt and Company, New York, 2014.

Laursen, Charlotte. *War on Wildlife Crime: The Illegal Ivory Trade.* Lambret Academic Publishing, London, 2016.

Love Nuwer, Rachel. *Poached: Inside the Dark World of Wildlife Trafficking.* Scribe, London, 2018.

Pinnock, Don in Bell, Collin. *The Last Elephants.* Smithsonian Books, Washington, 2018.

Rademeyer, Julian. *Killing for Profit: Exposing the Illegal Rhino Horn Trade.* Penguin Random House, Cape Town, 2012.

Sommerville, Keith. *Ivory: Power and Poaching in Africa.* Hurst & Company, London, 2019.

Svensson, Patrik. *The Book of Eels: Our Enduring Fascination with the Most Mysterious Creature in the Natural World.* Picador, London, 2020.

INDEX

INDEX

Mutai, Zachary, 3, 8, 47, 49, 99,
    116, 127, 133, 140
Mutisya, Samuel, 164

Nabire the northern white rhino,
    12, 13, 87, 108, 118, 182–183,
    190
Nasima the northern white rhino,
    12, 182
Ngulu, Stephen, 133, 141
Njenga, Abraham, 154, 155, 171,
    172
Nola the northern white rhino,
    13, 82
Northern white rhino
    (*Ceratotherium simum cottoni*):
    BioRescue project, 6, 8, 9,
    29, 77, 86, 88, 92, 96, 100,
    103, 107, 110, 119, 120, 134,
    136–137, 139–142, 182, 186–
    188, 191, 193, 195, 200, 201,
    203; difference from southern
    white rhino, 3; embryos, 110;
    Garamba National Park, 3,
    11–13, 17–27, 29–30, 32,
    91, 92; last in the wild, 205;
    natural habitat, 5; pregnancy,
    90, 111; relocation, 123;
    reproduction in captivity, 1,
    2, 12, 182, 184; sexual cycle,
    109

Ol Pejeta Conservancy: black
    rhinos, 37, 52; business
    model, 37, 174; cattle

farming (livestock program),
    41; Dog (K9) Unit, 164,
    168; Mutai, Zachary, 3, 8,
    47, 49, 99, 116, 127, 133,
    140; northern white rhino
    conservation project, 35;
    private reserve, 44, 165,
    166; rangers, 166; rhino
    cemetery, 33–57; southern
    white rhinos, 90; tourism,
    36; Vigne, Richard, 34–36,
    37–41, 171
oocytes (immature eggs):
    collection, 81, 89, 90, 103,
    141; fertilization, 7, 87, 182;
    from northern white rhino,
    76, 190; from southern white
    rhino, 77; maturation, 103,
    182; stem cell technologies,
    114, 137, 139; organized
    crime syndicates, 4, 62, 157;
    Owuan the southern white
    rhino, 54, 90, 133, 139–140,
    141

pangolin, 59, 66
pigs: biomedical research, 104,
    107, 113; genetically
    modified, 113
poaching: corruption, 6, 65, 67,
    68, 69, 156, 177; elephant
    poaching, 69; fight against,
    73, 166, 169; in Africa,
    58, 69, 156; organized
    crime (international crime

222

13, 31, 62, 87, 88, 90, 108,
116, 117, 118, 120, 121, 124,
125, 126, 127, 129, 140, 182,
201; Ol Pejeta Conservancy,
2, 6, 7, 8, 12–13, 24–25,
29–32, 33–41, 44–45, 46, 47,
49, 51, 53, 55, 56, 76, 82, 87,
90, 92, 94, 95, 98, 100, 108,
109, 110, 111, 117, 120, 123,
127, 133, 138, 139, 141, 147,
154–155, 163, 164, 166–167,
168, 171, 172, 182–183, 201;
Owuan, 54, 90, 133, 139–
140, 141; Southern white
rhino embryos, 77, 110, 111;
surrogate mothers, 7, 8, 10,
77, 86, 89, 90, 110, 127, 139,
140, 141, 195; Tauwo, 2, 40,
48–49, 54, 139, 140, 163
Stejskal, Jan, 119, 120, 122–124,
126, 129, 130–131
stem cells: cell reprogramming,
185; embryonic stem cells,
104–105, 184; induced
pluripotent stem cells, 89,
182, 189–193; primordial
germ cells, 190, 192;
research, 7, 113, 186, 191;
technologies, 7, 81, 89, 94,
114, 137, 139, 141, 183, 184,
185, 186, 193, 195, 204
Sudan: civil war, 14, 17, 21, 23;
Darfur, 14, 23, 68; Janjaweed,
23, 68; Shambe National

Park, 18, 49; South Sudan, 5,
12, 23, 43
Sudan (the rhino): death, 87, 116;
Dvůr Králové Zoo, 1, 2, 6,
7, 12, 13, 31, 62, 87, 88, 90,
108, 116, 117, 118, 120, 121,
124, 125, 126, 127, 129, 140,
182, 201; Mutai, Zachary 3,
8, 47, 49, 99, 116, 127, 133,
140; sperm, 109; the last
northern white rhino male, 3,
13, 39, 49, 55, 140, 182; Ol
Pejeta Conservancy, 2, 6, 7, 8,
12–13, 24–25, 29–32, 33–41,
44–45, 46, 47, 49, 51, 53, 55,
56, 76, 82, 87, 90, 92, 94, 95,
98, 100, 108, 109, 110, 111,
117, 120, 123, 127, 133, 138,
139, 141, 147, 154–155, 163,
164, 166–167, 168, 171, 172,
182–183, 201
Sumatran rhinos, 60, 64, 65, 91,
108, 114; earthquake and
tsunami in 2004, 94
Suni the northern white rhino, 2,
6, 12, 30, 31, 34, 87, 99, 109,
110, 116, 118, 182

Tanzania, 31, 42, 67, 68, 92;
Amboseli National Park,
31
Tauwo the southern white rhino,
2, 40, 48–49, 54, 139, 140,
163

About the Authors

Boštjan Videmšek (1975) is a long-term war correspondent. He has covered all major conflicts in the last 25 years (Afghanistan, Iraq, Syria, Darfur, Gaza, Somalia, DR Congo, Libya, Ukraine, Kosovo, . .). During the last five years he focused his work on the consequences of climate change—looking for solutions. His work has been published by the *New York Times*, Der Spiegel, Stern, *Boston Globe*, CNN, BBC, *National Geographic*, El Figaro, Forbes, Aftenposten, Vice, GEO, Geographical, *Sydney Morning Herald*, Mo, Revolve Magazine, El Periodico, El Diairo, Gulf News, We Demain, Caravan, Publico, *Sierra Magazine*, Politico, Middle East Eye, . . . Videmšek is the author of seven books (three translated in English, one in German) about modern conflicts and migration. His book *Plan B: How Not to Lose Hope in the Times of Climate Crisis*, a book about the most promising practices in the fight with climate change, has been named a book of the year 2020 in Slovenia. Videmšek, also an author of two theater plays and an ultra-marathon runner, is a recipient of many international and national awards. He has been chosen as one of the European Young Leaders.

Maja Prijatelj Videmšek (1979) is a leading Slovenian environmental journalist, focusing on food system, animal rights, waste management, renewable energy, endangered species, climate change. . . . She writes for the Slovenian daily DELO. Her work has been published by many international media outlets, including the *Boston Globe*, *Christian Science Monitor*, *National Geographic China*, Das Magazin, Haaretz, and *Hindu Times*.